THE GREAT YARMOUTH HERRING INDUSTRY

THE GREAT YARMOUTH HERRING INDUSTRY

COLIN TOOKE

Frontispiece: A scene that evokes memories of the autumn herring season in Great Yarmouth, when the girls from Scotland were as much a part of the local scene as the boats and the strong aroma from the curing houses These two, early twentieth century Scots fisher-girls, although off duty, are still in their aprons and boots and with their fingers bandaged.

First published 2006 by Tempus Publishing

Reprinted in 2008 by
The History Press
The Mill, Brimscombe Port,
Stroud, Gloucestershire, GL5 2QG
www.thehistorypress.co.uk

Reprinted 2014

© Colin Tooke, 2006

The right of Colin Tooke to be identified as the Author of this work has been asserted in accordance with the Copyrights, Designs and Patents Act 1988.

All rights reserved. No part of this book may be reprinted or reproduced or utilised in any form or by any electronic, mechanical or other means, now known or hereafter invented, including photocopying and recording, or in any information storage or retrieval system, without the permission in writing from the Publishers.

British Library Cataloguing in Publication Data.
A catalogue record for this book is available from the British Library.

ISBN 978 0 7524 3760 6

Typesetting and origination by
Tempus Publishing Limited.
Printed in Great Britain.

Contents

	Acknowledgements	6
	Introduction	7
one	Sail to Steam	13
two	The Fishwharf	49
three	Scots Girls	81
four	Bloaters and Beatsters	101

Acknowledgements

This book would not have been possible without the help of many people. I would like to thank James Steward and Joe Carr of the Norfolk Museum Service, Time & Tide Museum for help and access to the Maritime Library, and Mr J. Fisk, keeper of the Bloomfield Archive. Peter Allard has given me the benefit of his knowledge on the fishing industry, as well as assistance with images from his collection and advice on the text. I thank Mike Kelly of HS Fishing 2000 Ltd for allowing me access to his curing works and Paul Williment for allowing access to his Gorleston net chamber. Lowestoft author Malcolm White has been generous in allowing me to use photographs from his collection, and ASCO Ltd allowed access to their quayside depot. Other photographs are reproduced by kind permission of Archant Ltd, John Read, Rosie Shipp and Liz Hunter. As always my wife Jan has been invaluable in helping to sort the images, and in providing encouragement to complete the book.

Photograph credits:
Norfolk Museum Service, Time & Tide Maritime Library: 19B, 34B, 36T, 36B, 37T, 37B, 58B, 61T, 92B, 93T, 111T, 118T, 120T, 127T.
The Bloomfield Archive: 27T, 29B, 32T, 32B, 33T, 33B, 35B, 72T, 113T, 113B.
Malcolm White: 18B, 20T, 30T, 30B, 38T, 40T, 40B, 44T, 44B, 45T, 46T, 68T, 79B, 97B, 112B, 116B, 121T, 127B.
Peter Allard: 22T, 22B, 23T, 23B, 26B, 31T, 43T, 56T, 56B, 59, 62T, 70T, 70B, 74B, 75, 80B, 88B, 89T, 98B, 98B, 117, 125B.
John Read: 28T, 28B, 89B, 116T.
© Archant: 38B, 47B, 72B, 73T, 77, 79T, 80T, 88T, 93B, 94, 100B, 103, 104T, 104B, 106T, 106B, 112T.
Liz Hunter: 84B.
Rosie Shipp: 29T, 118B, 120B.
107B by courtesy of HS Fishing 2000 Ltd.

Introduction

Of all the fish in the sea, Herring is King.
Once upon a time all the inhabitants of the deep gathered to choose a king.
The fluke occupied too much time in putting on his red spots and did not arrive until after the
election, by which time the herring had been made King. The fluke curled his mouth on
one side and said 'a simple fish like the herring king of the sea' and his mouth
has been on one side ever since.

The town of Yarmouth, which became Great Yarmouth in 1272 under a charter issued by Henry III, owes its existence to the herring, a fact symbolised in the first town coat-of-arms, which depicts three silver herring on a blue shield. The seas off this part of the British coastline teemed with herring and, as the sandbank at the mouth of the Yare emerged in Anglo-Saxon times, it attracted fishermen from the neighbouring village of Gorleston and from further afield. Here, on the barren sandbank, they built huts from driftwood, dried their nets, and sold their catch during the autumn fishing season, and it was from this humble beginning that the town of Yarmouth rapidly developed. Increasing numbers of merchants came to barter their goods in return for the highly valued herring. The herring, Latin name *Clupea Harengus*, arrived in the seas off this coast in vast numbers as part of its annual southern migration, and was caught in large quantities to be sold to all parts of this country as well as to the Low Countries and the Mediterranean area.

Little is known of the industry's early history but it was flourishing by the eleventh century. The first documentary evidence of the settlement that was to become Great Yarmouth, and of the importance of the herring, is to be found in the Domesday Book of 1086. At this time there was no Lord of the Manor, the town was owned by the king and all taxes were payable to the king. As the town became more prosperous, local merchants suffered as they were not allowed to trade with other towns or markets. However, this situation changed following the first charter granted to the town in 1208 by King John, and the town became a 'free borough', able to govern itself and trade freely and independently. The consumption of all types of fish in England increased rapidly, and between 600AD and 1600AD there was a fourfold increase in the importance of herring in the diet. It kept many people alive during the heavy population increase of the thirteenth century, when other food was scarce. Fish also provided a large part of medieval protein intake, with all social ranks consuming substantial quantities. Its importance was emphasised by compulsory fish days and by Lent. Religious reforms demanding abstinence from meat led to new food sources from the sea being exploited. The open Denes, outside the medieval walled town of Yarmouth, became the site for a gathering of merchants and fishermen from many parts of Europe, a gathering which became known as the Free Fair, one of the greatest and most important fairs of medieval England. These international gatherings lasted for forty days, from Michaelmas (29 September) to Martinmas (11 November), and among the visiting fishermen were many from the Cinque Ports, a group of towns on the south coast. The Cinque Ports consisted of Hastings, Romney, Hythe, Dover and Sandwich (with Rye and Winchelsea added later). Because of their geographic position these ports were deemed to be the most important in the kingdom and were bound by the king to provide ships and sailors to guard the southern coast against French invaders. In return they were granted commercial privileges, of which the most important was access for the area's fishermen to the herring-rich seas off Yarmouth. They were also allowed to dry their nets on the Yarmouth Denes and draw their boats up on the sands. The bailiffs of the Cinque Ports, known as the Portsmen, were empowered by a charter to visit Yarmouth for the duration of the Free Fair, to administer law and order and to see fair play, a system that led to much resentment and dispute with the local population, both on land and at sea.

It was during the fourteenth century that Great Yarmouth reached the peak of its economic prosperity, becoming one of the most important provincial towns in the country. Fish had become established as one of the principal items in the medieval economy, and the herring was an important part of the diet of people from all walks of life. Both in this country and abroad, the herring had become an essential high protein food. The Royal household, abbeys and monasteries all purchased herring from Yarmouth in large quantities, some religious establishments having their own warehouses in the town for the storing of fish. Until Henry VIII's religious reforms in the sixteenth century, large quantities of herring were consumed during the period of Lent.

In 1344 the town possessed 250 fishing boats, and the total number of vessels in the port exceeded that of any other port in England, with the exception of London. Three years later the town was able to supply Edward III with forty-three ships and over one thousand mariners for the battle of Sluys, more than all the Cinque Ports combined. In recognition of this, the King allowed the town to halve their arms of three herring with the three lions of the royal arms, forming the coat of arms that is used today. The fourteenth century also saw the introduction of the 'red herring', a fish that has been cured with salt and then dried and coloured by the heat and smoke from a fire.

In 1357 the Statute of Herring was authorised by Parliament and this brought the industry under a certain amount of government control, regulating the price and defining the measure on which herring sales were based. The basic measure was the 'warp', which was four herring, simply two fish being picked up in each hand and thrown into a basket. Thirty warps (120 fish) made one 'hundred', and ten hundred made one 'cran'. Ten cran or 12,000 fish made one 'last', which weighed roughly two tons. In later years, as the size of the herring diminished, the 'long hundred' was introduced, this being 33 warps or 132 fish. During the following centuries many more Acts were introduced to control the sale and catching of herring.

For many years rents were paid in herring, the provost appointed by Henry I to govern the town received a rent of 10,000 herring and, from 1362, a grant of a last of red herring was made annually to St George's Chapel, Windsor. One hundred herring were sent to the Sheriff of Norfolk whose duty it was to have them baked into pies, destined ultimately for the royal table. The Battle of Herrings became one of the great victories during the Hundred Years War with France. In 1429 Sir John Fastolfe, the builder of Caister Castle, with a small force of men and 500 wagons of salted herring, attempted to relieve the besieged town of Orleans. He was attacked by 3,000 French soldiers and, by forming his wagons of fish into a circle, he enabled his archers to drive off the enemy, allowing him to enter the town and relieve the English army.

Throughout the fifteenth and sixteenth centuries the Dutch fishing fleet was expanding, extending their fishing in the North Sea and English Channel as the size of the English fleet decreased. In 1610 the government took steps to protect the home industry and tolls were levied on foreign fishermen visiting the Free Fair. The fair began to decline, partly due to the introduction of a policy by Charles I prohibiting Dutch fishermen from entering English waters. This policy was continued by Cromwell, who went a step further by declaring war on the Dutch, a great blow to the prosperity of the town. From 1662 the annual visits by the Portsmen were discontinued. In 1671 when Charles II visited the town, following his restoration to the monarchy, he was presented with a gift of four herring made from solid gold.

For many years the town derived considerable revenue on the sale of herring by what was known as 'heyning money', the difference between the market price and a price fixed by the Corporation known as the 'tide price', an artificial price set by a group of freemen sitting in the Heyning Chamber at the Tolhouse. A prominent herring merchant, John Andrews, eventually stopped this illegal price fixing in 1709. Andrews was said to be the largest and richest herring merchant in Europe and lived in a house on South Quay, a building that is now the Custom House.

The Yarmouth fishing industry did not prosper again until the late eighteenth century when the Napoleonic Wars closed the Dutch ports to the rest of the world and the herring trade returned to England. It was now that the Scots became involved in fishing in a large way, following the herring migration south down the North Sea. The Free Fair no longer existed but, following the dawn of a new era of peace in Europe, the Dutch fishermen returned, bringing a new fair on the South Denes. This fair usually began on the last Sunday before Michaelmas, which was known as Dutch Sunday, and lasted for three or four days. People traveled from afar to witness the spectacle, where the Dutchmen sold pipes, toys, gingerbread and crystallised sugar sweets known as 'domino clumps'. The yellow-sailed Dutch fishing vessels, known as *schuyts*, were dressed overall, and made a colourful sight in the harbour, moored along the Ballast Quay, adjacent to the South Denes. These visits by the Dutch finished in the 1830s, but the Scots fishermen continued to make Yarmouth their main base during the autumn season. A first-class harbour close to the fishing grounds, and the shore based facilities for gutting, packing and exporting, enabled Yarmouth to retain its role as the leading herring port in the world. An Act of Parliament led to the construction of the Fishwharf, which opened for business on 5 November 1867. The use of South Quay for landing fish had become a nuisance to other port users and the special wharf, 2,251ft long, provided a solution. The development included a shed 750ft long with offices for the many companies involved in the herring trade and salt stores and refreshment rooms. In 1869 the port was home to 900 fishing boats of all descriptions, the industry employing 4,051 men and 531 boys. By 1907 it was estimated that 10,000 Scots fishermen, coopers, fisher girls and curers worked in the town during the autumn fishing season. The Scots developed new markets for the fish by exporting brine-cured, or pickled, herring to Germany and Russia, a trade that overtook the traditional export of smoked 'red' herring.

The end of the nineteenth century and the early years of the twentieth century saw the heyday of the industry, peaking in 1913. The town has over two miles of quay from the harbour mouth to the Haven Bridge, and in 1913 every foot was required. At least 1,163 boats were fishing from the port and between them they caught 2,488,140 hundredweight of fish, or over 800,000 cran. Six thousand Scots fisher girls were kept busy on the South Denes and at Gorleston gutting and packing the herring into barrels for export to Germany and Russia, while many tons found their way to the dozens of curing houses, where they were processed into various types of smoked fish, including bloaters and kippers. However, within a few years, the outbreak of the First World War and the Russian Revolution heralded the loss of lucrative foreign markets. For the duration of the war the Admiralty requisitioned many fishing boats including seventeen from Bloomfield's fleet, seven from Westmacott and four from Smith's Dock. The skippers and crews enrolled in the Royal Naval Reserve and the drifters, armed with six-pounder guns, served throughout the war as minesweepers and escort vessels along the Eastern seaboard. Over four hundred of Bloomfield's staff joined the colours. Fishing continued, on a much reduced scale, for the duration of the war.

The export trade did not return after the war, and many more countries now had their own fishing fleets. Over-fishing was resulting in smaller catches, and in 1933 the Herring Industry Board was set up to reorganise, develop and stimulate the industry. Catches continued to decline throughout the 1930s until, in 1938, the Board applied severe restrictions to the industry, from catching to curing. The average price of herring fell to the lowest it had ever been; the problem being that the catching power of the fleet greatly exceeded the limited markets available. When their voyage ended in mid-November, the Scottish fishermen declared that the season had been the most disastrous they had ever experienced. The fleet operating from Yarmouth that year totalled 494 boats, made up of 357 Scots, 88 Yarmouth and 49 Lowestoft drifters. Most of the Scottish boats came from Banff, Buckie, Fraserburgh, Inverness, Lerwick, Peterhead and Wick.

In 1928 a competition known as the D'Arcy Cooper Challenge Cup was introduced for vessels of the Bloomfield fleet. This annual competition was divided into two groups; one for steel-hulled boats and the other for wooden-hulled boats, and was for the most successful boat in the fleet over a period of 20½ working weeks. Suspended in 1939, the competition returned in 1947 and continued until 1962 as the Lady Cooper Cup. Another competition began in 1936, the annual Prunier Trophy. This was given by Madam Prunier who had opened her famous seafood restaurant in St James, Piccadilly, London in 1932, a restaurant that served exclusive seafood to the cream of London society. The trophy, made from Purbeck marble and sculptured by Charles Sykes, was awarded to the boat from Yarmouth or Lowestoft catching the most herring with one shot in one night. The skipper of the winning vessel was presented with a weather vane to be fixed to the mast. The last year the trophy was awarded to a Yarmouth boat was 1962, to the *Ocean Starlight*. The Prunier Trophy ended in 1966.

The outbreak of the Second World War brought the industry to a standstill, the boats once more requisitioned for war work. The port now became an important naval base, and the Admiralty requisitioned many buildings associated with the fishing industry. Fishing was able to resume in 1946, and in the 1950s the industry showed a slight recovery. By now, however, the North Sea was being fished indiscriminately by many countries. Trawlers were taking the spawning fish, to be used for oil, animal meal and fertilizer, and fish stocks rapidly declined. The home markets for herring also declined as the housewife's choice of new foods rapidly expanded. Birds Eye had begun production in the town, and experimented with freezing herring, but this was not popular. White fish became more popular than herring, leading to the introduction of the fish finger in 1955, which was developed at the Birds Eye Great Yarmouth factory. On 23 October 1950 at least three drifters landed catches exceeding 200 cran, a red-letter day for the industry in the difficult post war years. The top catch was by the *Wydale*, a catch of 250 cran which earned skipper Alfred 'Mabby' Brown of Caister the Prunier Trophy that year. The other boats with large catches that day were the *Ocean Spray* (YH 264) and a Peterhead boat, the *Fumerole*. Catches continued to increase and, on 21 October 1953, two boats, the *Ocean Lifebuoy* (YH 29) and the *Fruitful Bough* of Peterhead both landed catches of over 300 cran each. In 1955 the fishing fleet numbered over 150 boats landing 62,000 cran, or over 60 million fish. This, however, was less than half the catch of the previous year. A decline had begun, a decline that was to continue over the next decade.

In 1957 the English and Scots drifters briefly broke off from their autumn fishing to take part in the first review of the herring fleet. On board HMS *Wave*, the Commander-in-Chief The Nore, Admiral Sir Frederick Parham, took the salute as 170 fishing boats, drifters and longshore boats sailed past, watched by thousands of spectators on shore. That year only 111 drifters had fished from the port, catching only 8 per cent of the record 1913 catch. These figures continued to fall and, in 1963, the last vessel of the Bloomfield fleet was sold. In 1968 only five Scots boats fished from the port, and the following year there were none: the Yarmouth fishery had finished. The last local steam drifter to work out of Yarmouth was the *Wydale*. Today, the Scottish boats, fishing off the West Coast, catch many of the herring now sold in this country, while others are imported from Norway.

Apart from the ship builders, drifter owners and crews, many thousands of people were employed in the shore-based trades associated with the herring industry, both local people and a large seasonal influx from Scotland. The curing houses and yards, for hundreds of years tucked away in the Rows but by Victorian times expanding into new developments on the South Denes, cured the herring for markets both home and abroad. Among the many different types of cure were the red herrings and the famous Yarmouth bloaters and kippers. The pickling plots packed thousands of barrels of herring, packed in salt for export, while other factories canned the herring in oil or tomato sauce, or made bloater paste. The industry could not have operated without the almost unending list of trades allied to it, trades such as sailmakers, coopers, the beatsters and ransackers looking after the nets, the coal, salt and ice suppliers, the rope and twine spinners and the ship chandlers.

Now the drifters have gone, the Scots girls no longer make their annual journey south, and the curing houses lay empty. Little remains of the industry's legacy, and it will be difficult for future generations to comprehend the crucial role fishing has played in the economy of Great Yarmouth and the life of its inhabitants. The Fishwharf area was taken over in the late 1960s by the new oil and gas exploration companies and new freight services, and this led to the demolition or re-use of buildings once used by the fishing industry. The Yarmouth herring industry had disappeared, except in memories and photographs. It is an industry now relegated to the museums.

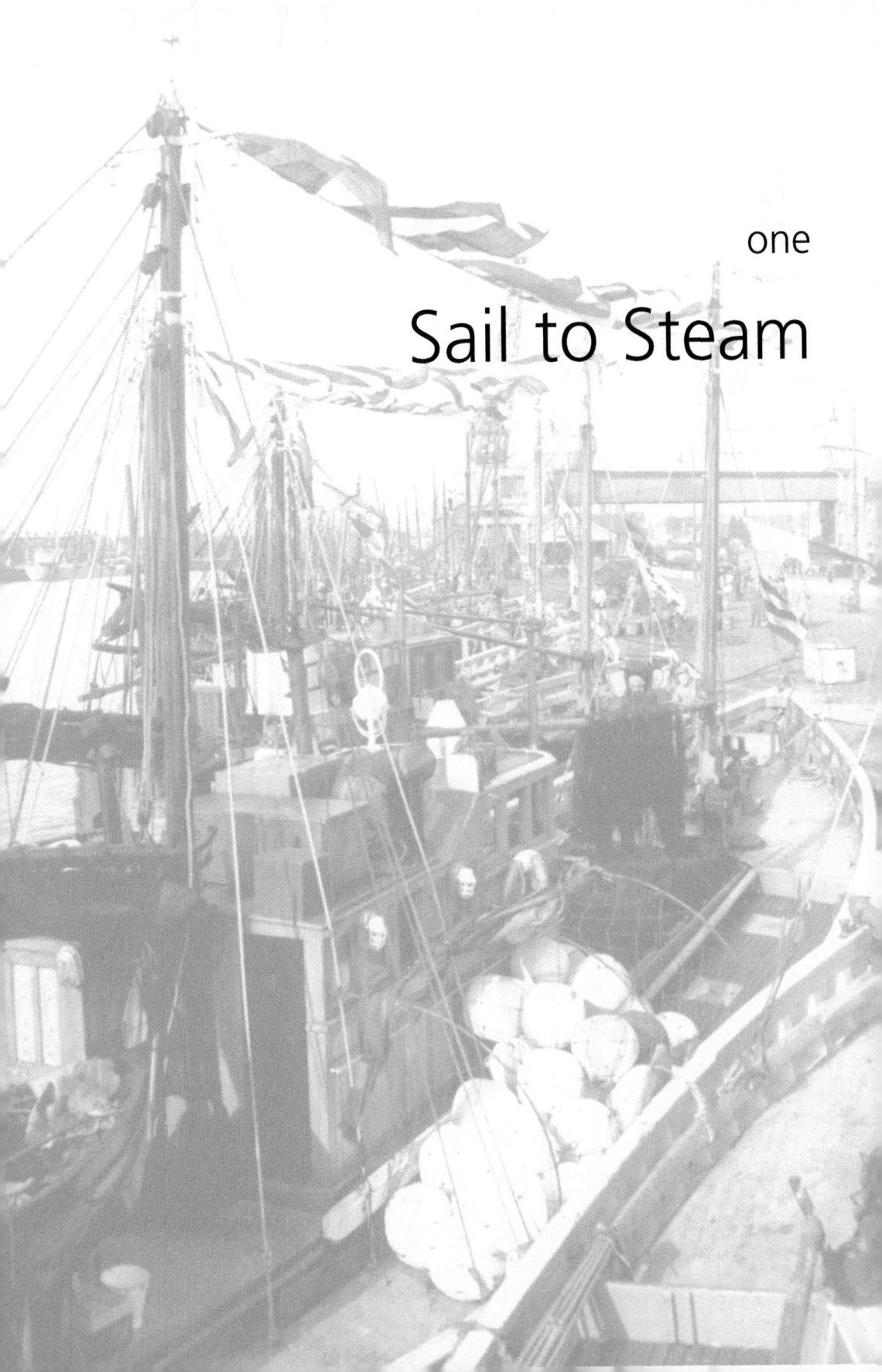

one
Sail to Steam

Throughout the nineteenth century, the vessels fishing out of Yarmouth were mostly the local sailing luggers and smacks, and the fifies and zulus of the visiting Scottish fleet. In 1853 the Yarmouth lugger *Perseverance* caught 126 last, or one and a half million fish, many of which had to be dumped overboard as the boat was unable to carry such a large catch back to port. In the 1860s, the luggers often remained at sea for up to six days, salting the catch after each haul. Each boat could carry fifty last of herring and six tons of salt, a last being 13,200 fish weighing two tons. Towards the end of the century great changes began to take place in the industry as sail gave way to steam. Horatio and Ernest Fenner, pioneers of steam, brought the first locally owned steam drifters to the port, the *Salamander* (YH 411) and the *Puffin* (YH 414). In 1896 the Fenner brothers placed an order at Beeching's yard for the first two steam drifters to be built at Yarmouth, the *Milton* (YH 236) and the *Queenborough* (YH 238). In August 1899 the first steel-hulled steam drifter to be built at Yarmouth, the *Claudian* (YH 380), was completed by Fellows & Co. at Southtown, also for the Fenners. Another pioneer in drifters was James Pitchers Junior, who bought his first steam vessel in 1897 and was to become the largest single owner of steam drifters at the port.

In 1900 four steel-hulled steam drifters, built on the Tyne, arrived in Yarmouth and were named, rather unromantically, *One* (YH 463), *Two* (YH 473), *Three* (YH 478) and *Four* (YH 480). These were the first boats in the fleet of the Smith's Dock Trust Co. and became known as the Red Funnelled Fleet or the Numerical Fleet. Of the twenty-eight steam drifters registered in the port in 1901, sixteen belonged to the Smith's Dock fleet, another eighteen vessels being added the following year. This fleet operated successfully from the port until the First World War.

In 1907 a total of twenty-one new steam drifters were registered at the port, one of these being the *Ocean Gift* (YH 574), built for James Bloomfield and William Green. In 1911 Bloomfield's Ltd became a registered company and a second drifter was purchased, the *Ocean Pride* (YH 172). This was the beginning of a fleet which continued to grow, until by 1923 it numbered twenty-three vessels and the company had been incorporated into the vast empire of the Lever Brothers. In 1928, two new boats, the *D'Arcy Cooper* (YH 370) and *Hilda Cooper* (YH 392) were named after Mr D'Arcy Cooper, chairman of Lever Brothers, and his wife. The number of vessels in the fleet declined in the 1950s as the industry declined, the last vessel registered at Yarmouth to fly the Bloomfield flag being the *Ocean Trust* (YH 377) in 1957. In 1963 the last six vessels in the Ocean fleet were sold, signalling the end of the fishing industry in the town.

There were many individual owners fishing from the port in the early years of the twentieth century, but the majority of boats belonged to the fleet owners, the largest being Bloomfield's, Crown Steam Drifters Ltd, Eastick's, Gt. Yarmouth Steam Drifters Ltd, Horatio Fenner, Smith's Dock Ltd, Pitchers Ltd and Westmacott. Three shipyards in Yarmouth built drifters, those of Beeching, Crabtree and Fellows. Motor drifters began to appear from 1905, but their introduction was slow and steam drifters were being built in the Yarmouth shipyards up to 1930.

A three-masted Yarmouth herring lugger unloading near the Haven Bridge, a drawing by E.W. Cooke in 1829. This was before the Fishwharf had been built, and many boats unloaded their catch on the beach, near the Jetty, while others used the river and quays. At this time the letters YH were not included in the boat registration number. The Haven Bridge in the background, a wooden lifting bridge, was rebuilt in 1835.

The Yarmouth sailing drifter *Maud* (YH 872) sailing down river on her way to the open sea, past the undeveloped South Denes and the Nelson Column. This drifter was built in 1883 by Mack Bros at their Southtown yard and owned by W. Stanley. The drifter was broken up in November 1909.

The sailing drifter/trawler *Aid* (YH 697), built in 1877, being towed out of the harbour by a steam paddle tug c.1895. On the right is the South Pier with the distinctive coastguard lookout at the end. In front of the lookout can be seen one of the capstans that were used to winch sailing boats into harbour. There was a row of these capstans along the pier, which was also known as the Dutch Pier, and which was rebuilt in the 1960s.

Opposite above: The paddle tug *Reaper*, owned by Nicholson's Towage Company, pulls a group of six sailing drifters out of the harbour. The group includes the *Nell* (YH 868), *Orient* (YH 1053), and a Lowestoft boat, *Primrose* (LT 87). The tug was sunk in 1901 following a collision off Gorleston pier.

Opposite below: The *Flying Childers* steam paddle tug, seen here leaving harbour c.1895, was bought in November 1890 by E. Durrant to serve his fleet of smacks and was sold in 1896 when the fleet was dispersed. The tug was named after a famous racehorse of that name.

A group of Scottish fifies, wooden built sailing drifters, being towed down river on their way to the fishing grounds, followed by some steam drifters. The Scottish boats come from Kirkcaldy (KY), Banff (BF) and Berwick (BK). One tug would usually pull a group of five or six drifters at a time.

A line of steam drifters moored south of the Fishwharf. In the background are the grandstands of the South Denes racecourse. Among the boats seen here, from second right, are the Yarmouth drifters *Boy Fred* (YH 517), *Young John* (YH 479), *Paradox II* (YH 712) and *Boudicea* (YH 664).

The *Benbow* (YH 350) was built in 1885 at the Southtown shipyard of Edward Castle, a yard that later became Crabtrees. The boat was sold to Norway in 1900 after a short spell at Lowestoft.

The crew on board *Our Boys* (YH 347), c.1905. This was the last sailing drifter to be built at Yarmouth, launched in July 1903. The boat was lost on Gorleston beach on 6 May 1913.

Beeching's yard in 1902 with a range of vessels in various stages of construction. On the far right is the *Yare* (YH 703), built for Great Yarmouth Steam Drifters Ltd, and fitted with a Crabtree engine. Beeching's yard, on Southgates Road, had closed by 1919, and the Trawl Market was built on the site, now the Atlas Wharf. In the background the two towers are part of the Gas Works.

The 32-ton, wooden sailing drifter/trawler, the *Star of the East* (YH 90), was built at Southtown in 1857.

Right: A Beeching Bros advertisement of 1894. James Beeching, their father, was the founder of the firm and a prolific boat builder, winning a prize in 1851 for his self-righting lifeboat design.

Below: The *William Henry* (YH 770) under construction in 1877 at the Gorleston shipyard of H.A. Morris, a yard just south of Ice House Hill. In the nineteenth century there were many small shipyards along the banks of the river Yare, building a large variety of vessels.

Left: The Hull steam drifter *Fraserburgh* (H 950), built in 1907 and sold to Malta in 1947, had an engine made by the Yarmouth firm Crabtree Ltd. Although registered at Hull, the boat fished out of Yarmouth with a predominantly local crew.

Below: Shade of Evening (YH 676) was a 22-ton drifter built in 1866 for G. Hobbs of Caister, seen here lying at East Quay.

Steam drifter *Inter Nos* (YH 973) was built in Yarmouth in 1907 and is seen here going on her sea trials before being handed over to her owners. On the left is part of the grandstand of the South Denes racecourse.

Spring Flower (YH 735) returning from sea trials in 1902. When a boat was on sea trials many people, including relatives of the crew, were invited to join the trip, the reason for the unusually large number of people that can be seen aboard this boat, and aboard the boat in the picture above.

A view from the Gorleston side of the river shows the drifters moored along South Quay. In the background are some of the thousands of barrels stacked on the Denes, and to the right the Roundhouse, a fish sale ring.

Drifters moored bow-on to the quay, a system necessary when a large number of boats required a berth at the same time. At peak times at the Fishwharf, as soon as a boat was unloaded it was expected to move, and free up the limited space available.

Taken in 1928 from the top of the temporary Haven Bridge, this picture shows a variety of boats including wherries and barges moored near the Town Hall. The drifters are moored five abreast on the Yarmouth side and three and four abreast on the Southtown side, leaving only a narrow passage down the centre of the river.

Scottish motor drifters moored at Hall Quay in the 1950s. These drifters were of wooden construction and a distinctive shape, built on the east coast of Scotland. The Scottish fleet did not put to sea on Sundays and the number of boats in the port at the height of the season meant there was a shortage of space. Boats moored the length of the quay, from the Fishwharf right up to the Haven Bridge.

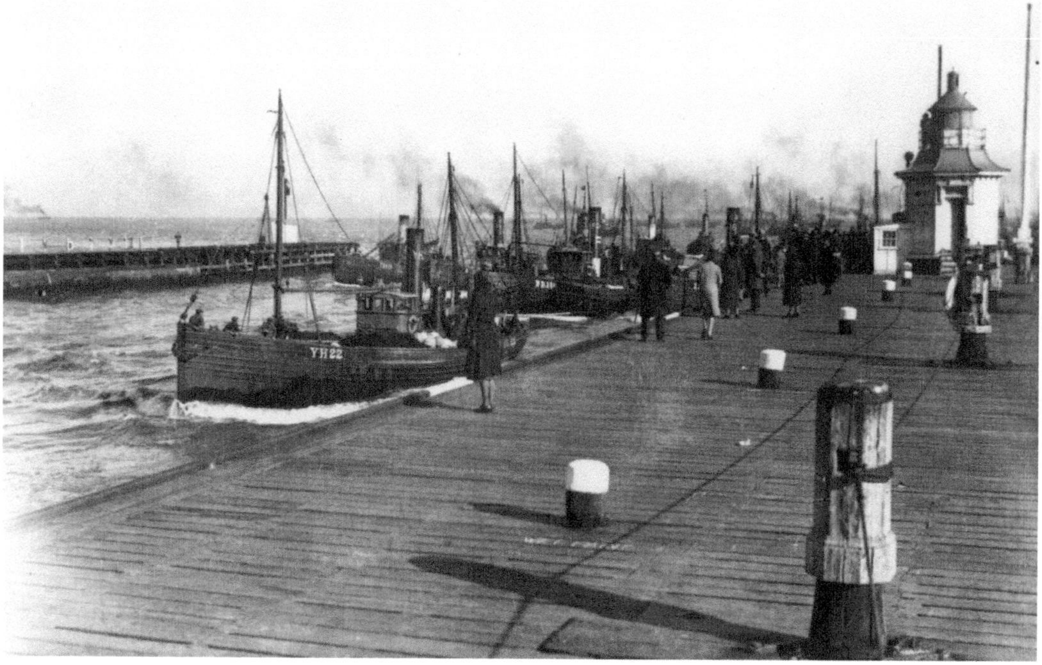
YH22

LUDHAM CASTLE
YARMOUTH

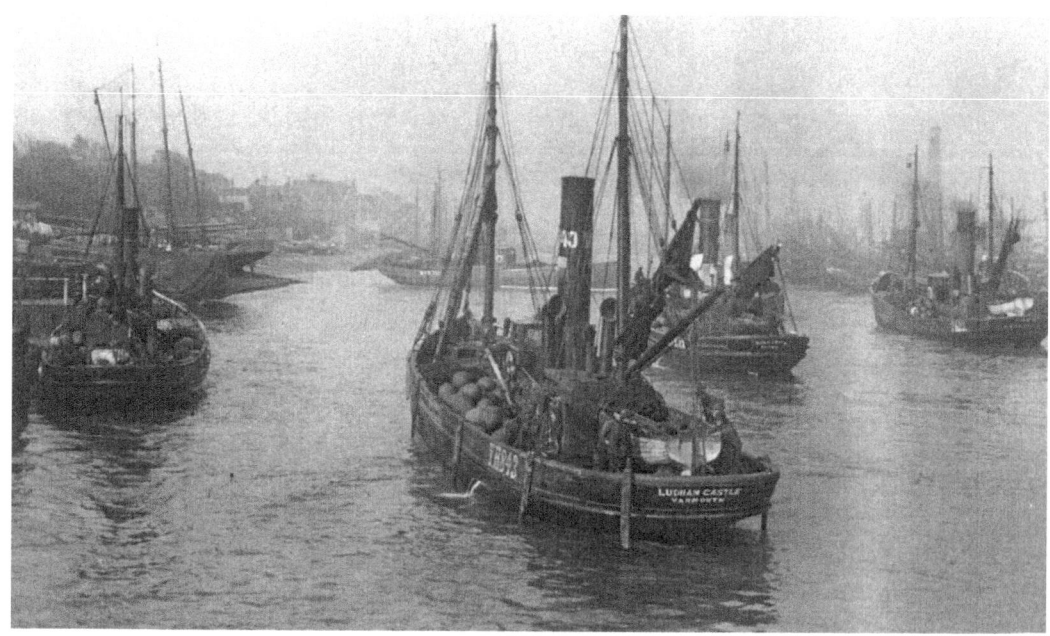

The skipper and crew of the *Ocean Retriever* (YH 307) in 1931, the winners of the D'Arcy Cooper Challenge Cup that year. This competition was for boats in the Bloomfield fleet, the winning boat receiving £100 to be divided among the crew. The *Ocean Retriever* was built at Lowestoft in 1912 and destroyed by mines in the Thames Estuary during war service in 1943. A drifter crew usually consisted of ten or eleven men, comprising the Skipper, Mate, Driver (or Engineer), Stoker, Hawesman (or Oarsman), Whaleman, Net Stower, First Younker (or Deckhand), Second Younker, sometimes a Third Younker and the Cook. Some of these unusual names were a legacy from the days of the old luggers and smacks. Each man had a specific job on board, and the earnings from the catch were divided between the owners of the boat, the skipper and the crew in a complicated system of shares.

Opposite above: A few onlookers on the Dutch Pier at Gorleston watch as the *East Holme* (YH 22) leads a large group of drifters back into the harbour. The early morning race back from the fishing grounds was important, as the first catches landed would command the highest market price.

Opposite below: The *Ludham Castle* (YH 843) was a 65-ton steam drifter built in 1904 in Devon for J.A. Gordon of Ludham. After service in the First World War she was sold to a French owner in 1920. The boat was lost off the French coast the following year.

The drifter *Grace* (YH 67) and crew. This boat, one of the fastest steam drifters at Yarmouth, had been built in 1884 and came to Yarmouth in 1898 when owned by James Pitchers Ltd. The tall thin funnels on boats such as this were known as 'Woodbine funnels', after the thin Woodbine cigarettes, popular at that time.

A few onlookers on the South Pier watching the boats come down river, leaving for the fishing grounds, c.1912. The boats would usually leave harbour in late afternoon, in time to reach the fishing grounds by early evening.

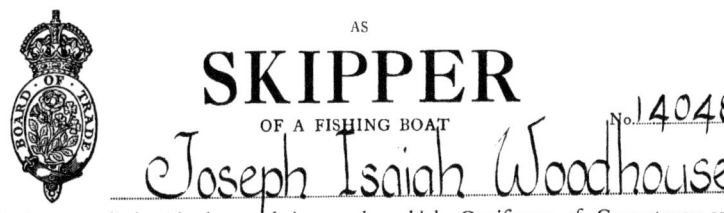

A 'Skipper's Ticket' as issued by the Board of Trade. The skipper had a great responsibility and in many cases could be the only literate person on board. It was up to the skipper to choose the spot to shoot the nets, a decision that was vital to the success of the trip. Joseph Woodhouse of Caister was skipper of the *Oak Apple* (YH 365), a steam drifter belonging to the Westmacott fleet.

A boat from the Bloomfield fleet, *Ocean Veteran* (YH 151), moored at the Fishwharf next to a drifter from Peterhead. The boat was originally a Scottish boat, the *Darda*, and was bought by Bloomfields in 1947 and renamed *Ocean Veteran*. The boat was broken up in 1954. The slogan 'Eat More Herrings' was that of the Herring Industry Board, the Government body set up in 1933 to promote and assist the industry.

The steam drifter *Hamnavoe* (YH 117) was built at Aberdeen in 1918. In 1953 she was owned by Paul Williment of Gorleston but then sold to George Newson and later renamed *Aramanth*. Newson's lorry is seen here loading nets onto the drifter that was eventually broken up in Holland in 1955.

Opposite above: Twenty-six (YH 678) was one of the Numerical Fleet owned by Smith's Dock Trust Ltd, which operated from 1900 until 1915, when the company was wound up. This boat was requisitioned by the Admiralty in 1915 and fitted with a six-pound AA gun and used as a minesweeper. In 1922 the boat became the *Carpe Diem* (LT 1207) and was broken up in 1935.

Opposite below: The *Monitor* was built in Yarmouth at Fellows' yard for a Scottish owner, her home port being Buckie. The boat is seen here brand new, before her first fishing trip and probably waiting to be handed over to her owners.

The drifters usually reached the fishing grounds and were ready to shoot their nets at dusk, the time when the herring 'swam' or rose to the surface to feed on the plankton that made up their diet. The nets were 'shot', and the boat allowed to drift with the tide, the crew then having time for a meal and rest until the skipper decided it was time to haul, usually sometime after midnight.

The warp, the rope running along the bottom of the nets to keep them down like a wall, was brought aboard with the help of the steam capstan and coiled in the rope-room by the cook. The other members of the crew then hauled the nets in over a wooden roller fixed along the bulwarks as seen here. The top of the nets was fixed to a head rope, corks and buoys keeping the line of nets upright in the water.

When the nets were on board the herring could be shaken from the mesh, where they had been caught by the gills, and shovelled with wooden shovels into the hold of the boat. The haul lasted without a break for at least five hours, a heavy catch often taking twice as long.

As soon as the catch was on board the skipper made full-speed for port and on the way the crew cleaned the nets and stacked them away. Up to one hundred nets could be used by each boat, a line that could stretch for two miles.

The drifter *Lydia Eva* (YH 89) ready for launching at Kings Lynn in 1930. Note the refreshments being served at the tables in the foreground. The hull was built by the Kings Lynn Slipway Company and then towed to Yarmouth to be fitted out and the engine installed at Crabtree's yard.

Built for Harry Eastick, one of the port's largest individual drifter owners, the boat was named after his daughter Lydia Eva. The fishing career of the *Lydia Eva* was short, lasting for only eight years. In 1938 she was requisitioned by the Admiralty and in 1947 renamed *Watchmoor*, working for the Royal Navy until 1969. The boat now belongs to a trust and is awaiting restoration at Lowestoft.

The *Ocean Hunter* (YH 296), a Bloomfield boat, berthed at South Quay in 1951. In the background is the Town Crane and the two rail-mounted cranes that belonged to the Port & Haven Commissioners. On the opposite side of the river are the wood yards of Jewson and Palgrave Brown.

The *Ocean Crest* (YH 207) was a motor drifter/trawler built in 1956 for the Bloomfield fleet. The boat later fished out of Lowestoft and in 1967 was sold to the University College of Swansea, starting a new career with their Oceanography Group, studying submarine geology.

The *Ocean Dawn* (YH 77) was built by Richards Ironworks at Lowestoft for Bloomfields in 1956. The first of a new class of motor-powered drifter/trawlers she was sold to Small & Co. (Lowestoft) in 1963 when the Bloomfield fleet was dispersed. Six years later the *Ocean Dawn* went to Scotland, re-registered at Kirkcaldy. This boat is still afloat and made a visit to Yarmouth in 2000 for the Maritime Festival.

Supper aboard the *Ocean Dawn* in the late 1950s. Accommodation was very limited for the ten-man crew. Note the wedge-shaped table designed to fit the restricted space and the wooden battens to hold the tableware in place during rough weather.

The *Ocean Starlight* (YH 61) was launched from Richards Ironworks at Lowestoft on 4 September 1952. This boat was destined to become one of the best-known fishing boats ever to work out of Yarmouth. She sailed on her maiden fishing trip on 8 October 1952.

The *Ocean Starlight* leaving harbour. On the evening of 9 November 1962 the *Ocean Starlight* put into Yarmouth with a catch of 294 cran of herring. This catch won the boat the Prunier Trophy that year, and also the Lady Cooper Cup as Bloomfield's most successful drifter of the year.

Scottish drifters, one from Buckie, moored at the Fishwharf in 1954. When the boats left for home in December the decks would be laden with furniture and other large items, purchased in the town and taken home for Christmas.

Fish being unloaded at a crowded Fishwharf. The fisherman is standing on the *Refraction* (YH 111), originally a Scottish boat built in Banff in 1919 and re-registered at Yarmouth for the Eastick family in 1953. Another Yarmouth boat is moored behind the *Refraction*.

The *Wydale* (YH 105) moored near the Fishwharf in 1958. Belonging to the Eastick fleet, this boat was the last operational steam drifter at Yarmouth and the last to fish from a United Kingdom port. Built in 1917 at John Chambers' Oulton Broad yard she left Yarmouth on 29 October 1961 for a breakers yard in Holland. On the masthead is the Prunier Trophy wind vane she won in 1950 (now on display at the Time & Tide Museum, Great Yarmouth). On the right of the picture can be seen the overhead coal conveyor which supplied the old power station that closed in 1959.

Looking across to Hall Quay the Scottish fishing boats are moored on both sides of the river, indicating it is a Sunday, when only English boats went to sea.

Some of the crew of the *Ocean Spray* (YH 264), a steam drifter built at Lowestoft in 1912. Moored behind is the *Thirty* (YH 695), one of the Smith's Dock numerical fleet. The scene is c.1913, the peak year of the industry.

The *Morning Star* (YH 479) was a steam drifter built in 1907 and transferred to Scotland in 1913. The following year the registration number was allocated to a new drifter built at Yarmouth, the *Young John*, that fished from the port until 1953. This picture shows the mizzen, the small aft sail that was used to steady the boat while the nets were shot, tied in a 'swallow tail'. This was a distinctive feature of Yarmouth boats when returning to port, a distinguishing sign to those watching from shore.

The crew of the sailing drifter *Emerald*, (YH 1014), wrecked on Scroby Sands and taken to the Sailors Home on 29 October 1913. Several drifters were wrecked off Yarmouth, the crews rescued by lifeboat and taken to the Sailors Home. After being a Maritime Museum for several years the building today has a very different role, that of a Tourist Information Office.

The wreck of the drifter *Tryphena* (YH 276). Returning to harbour in the worst gale of the year, she hit the pier and came ashore on the South Beach on 5 December 1929, the crew rescued by the Rocket Brigade. For many years a boiler was visible at low tide on the South Beach, a boiler which either came from this wreck or that of the *Olive*, another drifter wrecked in the same place on 27 October 1933.

A Peterhead drifter *Spes Aurea* moored near the Lower Ferry. The ferryboat has just arrived from the Southtown side with a full compliment of passengers, most of them appearing to be Scottish fishermen in their 'Sunday best'. On the left can be seen the Gorleston Gas Works.

The Lower Ferry leaving the Southtown side of the river on a weekday. On the Yarmouth side a few drifters are moored at the Fishwharf and others further down river. In the background can be seen the Yarmouth Gas Works.

Two Yarmouth drifters, *Young Cliff* (YH 126) and *Harry Eastick* (YH 278), in Fellows No.2 dry dock at Southtown, both drifters part of the Eastick fleet. The *Young Cliff* spent the war years as a mine recovery vessel and was broken up in 1959 while the *Harry Eastick* was broken up in 1961 in Holland.

A busy time at Fellows with four drifters in the No.1 dry dock. This shipyard had started in 1825 and was taken over by Richards Ltd in 1970. Still in use today, it is now the only shipyard at Yarmouth.

Right: James Bloomfield came to Yarmouth in 1902 to work for the Smith's Dock Trust Company and in 1911 founded the firm of Bloomfields Ltd. During the First World War he was given the naval rank of Honorary Commander for his services with the Admiralty. Bloomfields was to become the largest firm in the fishing industry at Yarmouth. James Bloomfield died in 1922, aged 54, and is buried in the Catholic Cemetery, Caister Road, Great Yarmouth.

Below: The crew of the Scottish drifter *Fruitful Bough* pose for the camera after a successful trip. The sweaters, worn by all the fishermen, were knitted by the Scots fisher-girls.

The *Phyllis Mary* (YH 578) in the process of being broken up at Seago's yard in 1956. Moored behind is another drifter waiting to be broken up, the *Romany Rose* (YH 63). This boat had won the Prunier Trophy in 1946. This fate awaited many drifters in the late 1950s as the fish stocks in the North Sea, and the industry, declined.

The skipper and crew of the *Ocean Starlight*, Prunier Trophy winners in 1962, the last year that a Yarmouth boat won the trophy. Declining fish stocks and falling catches led to the trophy being discontinued four years later.

The mayor, Councillor John Birchenall, presents the skipper of the *Ocean Starlight*, Mr Stanley Hewitt, with the mast-top wind vane in 1962.

The Prunier Trophy, usually kept by the owners of the vessel, was made from Purbeck marble and depicted a hand rising from the waves clutching a herring. It was carved by the sculptor Charles Sykes. For many years its originator, Madame Prunier, a London restaurateur, presented the trophy, but in 1958 she handed it over to the Herring Industry Board. The trophy is today in the Lowestoft Maritime Museum.

This large model of a drifter, built to scale and named *Herring Catcher*, was entered as a float in the 1934 Yarmouth Carnival by the Herring Industry Board. After the carnival the float toured many of the principal towns in the country, as an advertisement to promote the industry and to encourage people to eat herring. It is seen here passing the open air swimming pool on the Marine Parade.

Another view of the carnival float, YH 1934. 'Eat More Herring' was the slogan of the Herring Industry Board. The *Herring Catcher* is seen here at the Wellington Pier, the 'crew' collecting money in their nets.

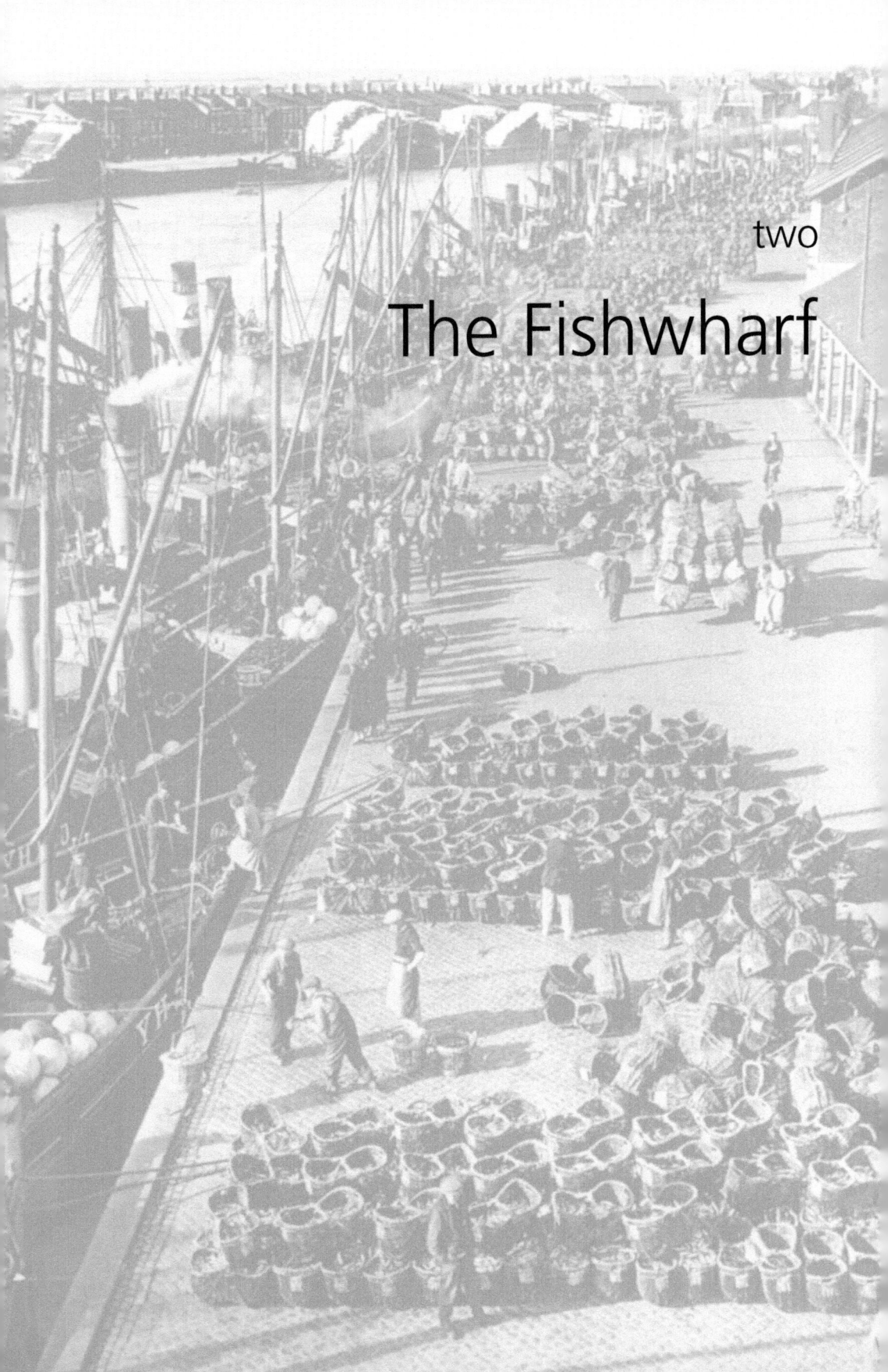

two

The Fishwharf

For hundreds of years the Yarmouth fishing boats unloaded their catches on the beach, near the Jetty. The boats grounded in shallow water and offloaded their catches directly on to the beach where they were auctioned. From here the herring were carted to the town's curing houses by troll carts, narrow horse-drawn vehicles that were able to negotiate their way through the narrow Rows to where many of the small curing houses were situated. In the early years of the nineteenth century the luggers began to unload their catches on the quaysides, in no particular place and to the annoyance of other users, particularly on the South Quay. As the number of fishing boats increased it became apparent that a special part of the quay had to be reserved for the industry and an Act of Parliament was sought to construct a Fishwharf. This opened for business on 5 November 1867, a wharf 2,251ft long with a covered fish market 750ft long. Other buildings, such as salt stores, an ice house and refreshment rooms were soon added. By 1902 there were twenty-one offices on the Fishwharf, catering for the fish salesmen, fishing companies, five railway companies and four banks.

A post office was built to deal with the thousands of telegrams sent during the herring season. The main post office in Regent Street received telegrams by teleprinter, and these were forwarded to the sub office on the Fishwharf by an underground pneumatic tube which, with a length of one and a half miles, was the longest such tube in the country. Up to 600 telegrams were sent every day between the two buildings, a journey that took 2 to 3 minutes. Early morning telegrams conveyed the market prices to customers in this country and abroad, while evening telegrams would report to boat owners about the day's catches and the earnings of each drifter. Telephones eventually replaced the telegrams, and the pneumatic tube was last used in 1953.

On 21 October 1930 the Fishwharf received a royal visitor, HRH the Prince of Wales, Master of the Merchant Navy and Fishing Fleets, who was in the town to open the new Haven Bridge. In the morning he inspected a drifter, the *Eastholme*, watched as part of the catch was unloaded, and then attended a herring sale. The Prince then visited Bremner & Low's fish-curing premises where he inspected the Scots girls at work. Following this he visited Bloomfield's net chambers and Sutton's curing works.

The catch was winched from the boat's hold on to the quayside in quarter-cran baskets, and then tipped into swills, one swill holding about 500 herring. The swill was a basket unique to the port of Yarmouth and used until 1960 when replaced by metal boxes. The loaded swills were taken to the covered market to be auctioned before being moved to the curing yards or pickling plots.

Herring measures:
4 herring = 1 warp
30 warps = 1 hundred
10 hundred = 1 cran
10 cran = 1 last

When the fish diminished in size the 'hundred' became 132 fish, a 'long hundred'.

Small wooden carts, known as troll carts, were used to carry fish landed on the beach into the town. These special carts, twelve feet long and three feet wide, were unique to Yarmouth and were designed to negotiate the narrow Rows of the old town, where all the curing houses were situated until the nineteenth-century development of the South Denes.

Before the Fishwharf was constructed most catches were landed on the beach near the Jetty and auctioned there before being carted to the town's curing houses. This engraving from 1846 shows the fish being brought ashore by small boats from the luggers. The swills are then transferred to the troll carts.

Left: Billy Mann was a well-known fish salesman, one of many who used the area at the shore end of the Jetty for their sales pitches, and the nearby Barking Smack pub as their office. Large quantities of mackerel as well as herring were landed near the Jetty, and in 1854 there were 84 boats involved in the mackerel fishery, which that year was worth £34,253.

Below: A busy scene at the Jetty in the 1860 fishing season, as captured in this watercolour by local artist W. Rowland. The original Jetty was built in 1560 and had a small crane at the end. It was rebuilt in 1710 and again in 1808 and lengthened twice in the nineteenth century.

Marine Parade in the late 1870s. The goat carts are waiting to take children for rides along the Parade. The buildings in the background are, from the left, the Bath Hotel annex, demolished in 1906 to make an open forecourt for the Hippodrome, (now the site of the Circus Circus amusements), the Fish Warehouse, also known as the Fish Station, used when large quantities of fish were landed and sold on the beach, the Telegraph Office, the Norfolk Hotel (now a modern pub called Rudey's) and the Sailors Home. The Golden Nugget amusements now stand on the site of the Fish Station and Telegraph Office. The tall lookout belonged to the Young Company of beachmen, the tallest of several lookouts to be found along the Marine Parade at that time.

The Jetty in the late nineteenth century. The mackerel fishery was an important part of the Yarmouth fishing industry, the season being from 12 May to 12 July. The mackerel, a very perishable fish, were sent to London by cutters and by rail. Swills were not used for mackerel, the fish being put in the large wooden tubs seen here. In 1856 many Scottish boats were involved in the mackerel fishery.

An undated engraving showing a busy scene as sailing smacks unload their catches at the Fishwharf, probably shortly after it had opened in November 1867. It was now no longer necessary to land all the fish on the beach and transport them into the town, although small quantities were still being landed on the beach into the 1880s.

Above: Swills of herring, unloaded from the sailing drifters, ready for auction in the Fishwharf sheds in 1898. No doubt the gentleman in the bowler hat is a fish salesman.

Right: Salmon & Son, a firm with a very appropriate name, had, as this advertisement of 1894 shows, extensive interests in the fishing industry. Salmon Road, off South Denes Road, was named after this firm.

SALMON & SON,

AUCTIONEERS,

Fish ✢ Salesmen,

SMACK OWNERS & SALT MERCHANTS.

Fire, Life & Accident Insurance Agents.

J. T. SALMON—Secretary to The Great Yarmouth Ice and Co-operative Company, Limited.

Office—
FISH WHARF,
GREAT ✱ YARMOUTH.

Left: Two well-known characters in the local fishing industry: Alf Brown, the fish curer on the left, and Mr R.H. Beazor, a fish merchant, discussing business on the Fishwharf in 1898. The firm of Beazor were fish merchants until the 1950s.

Below: A group of well-dressed Victorian fish buyers at an auction on the Fishwharf in 1898. The salesman is the gentleman with the bowler hat.

A troll cart outside the Fishwharf salt store of Norford Suffling in the 1880s. Suffling was another one of the large firms engaged in the fishing industry, and a road on the South Denes is named after them.

Herring for export being loaded onto a ship at South Quay. After being packed into barrels with layers of salt, tons of herring were exported, many to Scandinavia, Poland and Gemany. In this 1930s scene the coal trucks wait in the sidings, ready to proceed to the Fishwharf to refuel the steam drifters.

This busy Fishwharf scene was captured on 21 October 1938, the empty swills waiting as the catches are unloaded. In the foreground the drifter unloading is the *Young Ernie* (YH 55). The lady on the quayside with one foot on the boat has no doubt found an opportunity for a brief chat with her husband before the boat puts to sea again. At times such as this when space was limited, each boat had to moor bow-on to the quay, adding to the difficulty in unloading.

Opposite above: Herring unloaded onto the Fishwharf ready for auction. Among the boats in the background is *Thirty* (YH 695), originally one of the Numerical Fleet of Smith's Dock Trust, a fleet in which the boats had numbers instead of names. In 1920 *Thirty* came into the ownership of John Hector Fuller of Yarmouth.

Opposite below: The catch of the Yarmouth steam drifter *Frons Olivae* (YH 217) being loaded on to motor transport, watched by a large group of onlookers in the 1930s. On the right a coal lorry awaits to refuel the drifter so she can put to sea again with minimum delay.

This catch is being loaded into Bloomfields barrels for transport to their curing houses. In the background is the Gorleston Gas Works, which was demolished in 1967.

By the late 1950s, when this picture was taken, there was still a small export trade in pickled herring to the Mediterranean. This was very small however, when compared with the huge trade to countries such as Russia in the peak years of the industry, before the First World War.

SMITH'S DOCK TRUST Co.
LIMITED.

GREAT YARMOUTH.

FISH SALESMEN

Coal & Cutch Merchants.

SHIP & ENGINE STORES.

Every Requirement for Fishing Vessels in Stock.

OFFICES:

Smith's Wharf, Gt. Yarmouth	Fish Quay, Lerwick.
Fish Wharf ,,	Fish Quay, Newlyn.
33, Sandhill, Newcastle-on-Tyne.	Fish Quay, Blyth.
17, Gracechurch St., London, E.C.	Fish Mart, Baltasound.
Albert Quay, Aberdeen.	Herring Market, Lowestoft.
Fish Wharf, Milford Haven.	14, Waveney Rd., Lowestoft.

TELEGRAPHIC ADDRESSES—
  Trust, Great Yarmouth.
  Smith's, Newcastle-on-Tyne.
  Ubication, London.
  Trust, Aberdeen.
  Trust, Milford Haven.
  Trust, Lerwick.
  Trust, Lowestoft.

NATIONAL TELEPHONES—
  256, Smith's Wharf, Gt. Yarmouth
  112, Fish Wharf   ,,
  7, Gorleston.
  2464, Newcastle-on-Tyne.
  2921, Avenue, London.
  260, Herring Market, Lowestoft.
  275, 14, Waveney Rd.   ,,

The Smith's Dock Trust operated a fleet of drifters, named *One* to *Thirty Eight*, and known as the Numerical Fleet, from 1901 until the First World War. The company had many other interests in the fishing industry, in many parts of the country, as can be seen from this advertisement of 1912.

The export wharf under construction, c.1910. Built to the south of the Fishwharf, on East Quay, this was built at a time when the export trade was at its peak and space at the Fishwharf itself was at a premium.

Dozens of small offices were provided at the Fishwharf for the fish salesmen, fishing companies, banks, railway companies and other firms who had an interest in the industry, such as the Scottish Freshing Co., pictured here.

Right: Parts of the Fishwharf were damaged in the first air raid on this country, 19 January 1915, when a German Zeppelin, *L3*, dropped several bombs on the town. Here the water tank, used to wash down the area, has been damaged.

Below: Another casualty of the air raid was the Fishwharf Refreshment Rooms, the Lacon's pub earlier known as the Fishwharf Stores and today as the Dolphin. The original pub had been built at the same time as the Fishwharf, but the decorative terracotta tiles, which have a nautical theme, were added during a rebuilding scheme in 1903.

The Fishwharf Post Office, damaged in the Zeppelin raid in January 1915. On the left is the office of the London & North Western Railway and on the right is the public telephone call box. The Post Office, which was only open during the fishing season, dealt with thousands of telegrams in the days before telephones became widespread.

The Coffee Tavern and Grocery Stores of Robert Redgraves, c.1910. Also in this building was the Ship Chandlery of Yarmouth Stores Ltd. In the background is the salt store of the Great Yarmouth Fish Selling Company, one of several salt stores at the Fishwharf.

Commission Agent.	
Salt Merchant.	☞ ESTABLISHED 1887. ☜
Cutch Merchant.	
Cotton Herring and Mackerel Net Manufacturer.	**Horatio Fenner,**
Herring Curer.	*AUCTIONEER and*
Proprietor of the Gt. Yarmouth Steam Tanning Works	FISH SALESMAN,
Agency for Paine's Red Flares & Drift Rockets.	**Great Yarmouth.**
Steam Ship Owner.	

Horatio Fenner's advertisement in 1894. Fenner was another of the large firms engaged in the fishing industry in Yarmouth, and today Fenner Road still exists in the South Denes area. Cutch was used to preserve and waterproof the nets in the 'tanning process'.

A busy quayside scene from the days of sail in the late nineteenth century. Sailing drifters are moored at the quayside, and the young lad resting on the barrow appears to be reading or making notes, perhaps a budding historian recording the industry!

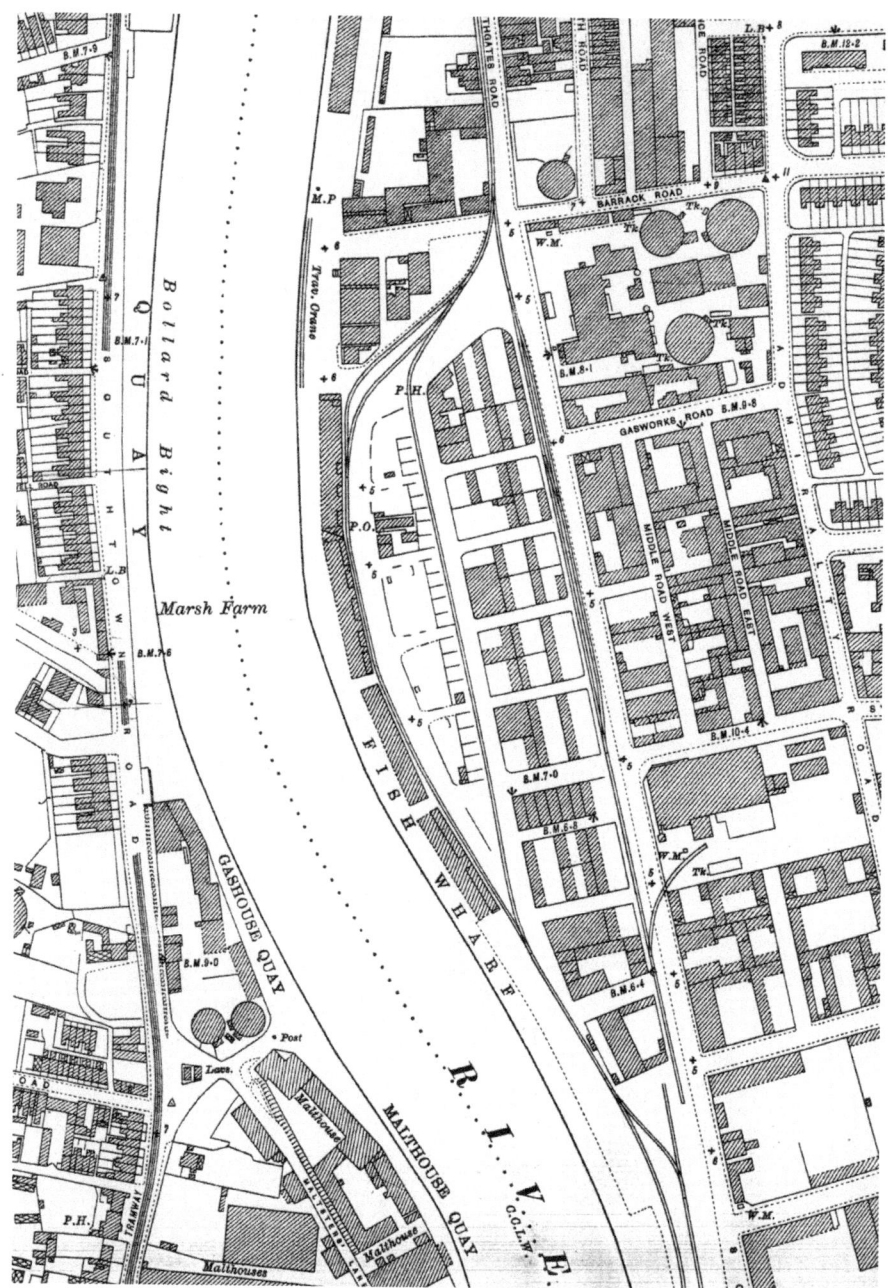

A section of the Ordnance Survey map of 1928 showing the Fishwharf area. Behind the main buildings are the 'pickling plots' and the Post Office. The extensive lines and sidings of the quay tramway, which connected the Fishwharf with the main railway network at Vauxhall Station, are shown. The line was laid in 1867 and used by the Great Eastern and M & GN railways. Large quantities of fish were moved by rail. Most of the buildings between the wharf and Admiralty Road were curing houses, net, ice or salt stores, or used for other aspects of the fishing industry. The siding on the right leads into the old power station, known as the Corporation Electricity Works.

HRH The Prince of Wales on board the drifter *Eastholme* (YH 22), 21 October 1930. The vessel was owned by John Carter and skippered by Mr A.H. Hubbard. The Prince, accompanied by Mr Russell Colman, chairman of the Port & Haven Commissioners, is watching as part of the 50-cran catch is unloaded.

Visitors were not discouraged from the Fishwharf. The couple in the centre of this group of onlookers appear to be holidaymakers, the fishing industry providing an added attraction to late summer visitors to the town.

This drifter has arrived in port with what would appear to be a good catch. Crew members pause for a photographer on the quayside, before the task of unloading begins.

Drifters from Peterhead unloading. Note the several different markings on the swills, indicating different firms. Although this postcard is postmarked 1962, the picture was taken several years earlier, probably in the 1950s.

A derrick, made up using the fore-mast and a pole known as the cran-pole, was used to lift the quarter-cran baskets from the boat's hold onto the quayside, a technique known as 'craning out'. In the background are Watney's Southtown maltings, now demolished.

The basket of herring is swung onto the quayside in the late 1950s, the drifter's winch used with the derrick to lift the baskets from the hold.

In the drifters hold the fish were loaded into the baskets using a metal, two handled, scoop known as a 'scutch'. This is part of the 294-cran catch of the *Ocean Starlight*, caught on the night of 9 November 1962, 40 miles E by N of Smith's Knoll.

Once on the quayside the quarter-cran baskets were tipped into the swills, one swill holding about 500 herring or half a cran. Twenty swills would make a last of herring, which weighed about two tons. Although this prodigious catch earned the *Ocean Starlight* the Prunier Trophy for 1962, it did not make much money, a little over £1,100. By the 1960s herring prices were at an all time low.

This aerial view of the Fishwharf, taken in 1952, shows the layout of the area, including the rail lines and the buildings associated with the fishing industry. In the centre of these buildings is the Corporation power station, and the overhead conveyor system used to take coal from the quayside into the station. At the bottom left is the Southtown malting complex of Watney Mann.

A 1950s scene on the Fishwharf. While the catch was being unloaded, the boats had to be replenished with fresh water, food, coal or fuel oil and replacement nets if required, ready to leave again for the fishing grounds a few hours later. There was no time for the crew to relax while in port.

Horse transport was preferred to motor transport on the Fishwharf, as it was more manoeuvrable in the congested area. Here, a Bloomfields cart, laden with empty swills ready for the next catch to be landed, is delivered to the quayside.

Nets being delivered from the net warehouse, after being mended and checked by the 'beatsters' and 'ransackers' which are to be seen later in this book. Again the horse and cart is the preferred method of transport.

Members of a drifter's crew take a breather, before preparing to sail again. As soon as the boat was unloaded it was necessary to prepare for the next voyage. Scottish crews and boats had a rest day on Sundays but the local boats usually worked a seven-day week.

A busy scene in October 1950. On the left the *Rose Hilda* (YH 73), a steam drifter/trawler, takes on coal from a lorry belonging to local coal merchant Bessey & Palmer. It appears that the gentleman in the bottom right of the picture, with the white bag, has managed to pick up a bag full of herring for tea. Among the other drifters unloading are the *Wydale* (YH 105), the *Achievable* (YH 92) and the *Animation* (YH 138).

Opposite above: This boat is unloading straight on to a company lorry in the 1950s with the owner no doubt standing on the quayside watching. It was the usual practice to allow any fish that fell from the baskets on to the quayside to be picked up by passers-by, many local families having free meals during the fishing season.

Opposite below: The fishing industry extended to the Gorleston side of the river; here empty barrels are being rolled ashore, ready for the Scots girls to fill with pickled herring. Although many barrels were made in the town, the huge quantity required exceeded the local supply, and many were imported.

5196 Herring Season, Gt. Yarmouth. Drifters Unloading their Catch.

In this 1958 view of the Fishwharf, the herring are being 'cranned out' straight on to the lorry, to be taken to the curing house. The lorry comes from Fraserburgh, many of the Scottish firms bringing men and transport to Yarmouth for the three-month fishing season. Several curing houses and pickling plots were owned by Scottish firms and stood empty for nine months of the year. The building in the background is today the only part of the Fishwharf left, this part of the quayside and the building now used as a depot by off-shore supply company ASCO Ltd. Firms supplying the oil and gas rigs in the North Sea replaced the fishing industry in the 1960s.

Opposite above: The *Clentanar* being loaded with barrels of pickled herring at the Export Wharf in the 1930s.

Opposite below: A drifter from the Scottish port of Buckie unloading her catch in the 1930s. In the background is the store of Norford Suffling. Many Scottish drifters were owned and crewed by one family.

Large quantities of fish were moved by rail, the quayside tramway linking the Fishwharf with the main railway network at Vauxhall Station. In 1886 the tramway was linked to the M&G.N. Yarmouth Beach Station by the Yarmouth Union Railway, a line that passed through a narrow gap beside the White Swan pub, known as the 'hole in the wall'. The train seen here has just entered the tramway from the Beach Station link.

The Rotunda, or Roundhouse, seen in several of the previous photographs, was built c.1912 as a herring sale ring, for fish landed by drifters who could not find a berth at the busy Fishwharf. The building is seen here in 1966 after being used by an engineering firm for several years. It was demolished the following year.

By 1960, when this picture was taken, baskets and swills had been replaced with metal boxes, the fish being unloaded straight onto the lorry. This prevented a lot of fish falling onto the quayside, but it looks as if the two boys on the right and the ladies on the left are hoping a few fish will fall from the boxes to provide a free tea.

The *Wydale* (YH 105) is the only boat unloading at the Fishwharf, *c.*1962, watched by a few hungry seagulls. Scenes such as this would not have been thought possible fifty years earlier, when boats had to wait for a free space before they could berth to unload their catch. The end of the industry is in sight.

In 1960 it was decided to replace the swills with metal boxes. That year, much to the delight of the local children, an enormous bonfire was built on the beach for 5 November. The fat from millions of herring that had soaked into the basketwork made the old swills highly combustible.

November 1962, and the few swills that remained on the Fishwharf are burnt in old oil drums. The end of an era.

three

Scots Girls

As the Scottish fleet of drifters arrived for the autumn season, so did the Scottish fisher-girls. These 'girls', a term used regardless of age, arrived in special trains from Aberdeen at Beach station. The girls came from many fishing communities in Scotland, including Banff, Buckie, Fraserburgh and Peterhead and the Shetland Islands. Before leaving Scotland the girls had signed contracts with specific curing firms, each curer having their own reserved carriages on the train. Each girl brought a trunk, referred to as a kist, containing her belongings. Their employers in the curing firms paid the rail fare and provided oilskin aprons and rubber boots. On arrival in Yarmouth, after a journey that had taken up to twenty-four hours, they went to their lodgings, most of which were at the south end of the town, close to the South Denes, where the curing houses and pickling plots were situated. Many went to Gorleston where they lodged in the area around Bells Road, the working areas being a place called Gut Meadow, off Pier Plain and along the riverside between the lifeboat station and Darby's Hard. The accommodation was sparse, the landladies only providing beds and a few basic pieces of furniture, such as a table and chairs; clothing was kept in the kist. One room would often be shared by at least six girls.

These highly-skilled girls were divided into crews, a crew consisting of one packer and two gutters, the packer effectively in charge of the crew. They worked mostly outside, working from a long sloping trough, known as a farlin, into which the lightly salted herring were tipped. With a small sharp knife the gutters would cut the throat of the herring, pull the gills and intestines out through the small slit, and throw the fish, according to size, into a tub. From the tubs the fish were packed into barrels, one crew able to fill about thirty barrels a day. One girl could gut up to sixty fish a minute, protecting her fingers with strips of cloth, known as cloots. The cloots gave some protection against cuts, salt sores and fish bones, but there were still several injuries to be treated by the dressing stations, of which the Red Cross dressing station in St Peter's Road was the most notable. The work was piecework, and the crews worked long days, anything from twelve to fifteen hours, working late into the night if necessary. At night they gutted by the light of Naphtha flares, a light that made the herring appear to glow silver. Scots girls never worked on a Sunday, many attending the local branch of the Church of Scotland on South Quay or the open-air services held on the Fishwharf.

At its height the fishing industry employed over 6,000 Scots girls in the town, gutting and packing the herring. When not working the girls were common sights walking along the quayside or around the town knitting fishermen's navy-blue jerseys or guernseys. As they walked along they talked to each other in an accent which local people found hard to understand.

The girls returned to their homes in Scotland just before Christmas, and they spent large amounts of money in the town on presents, many of them for children left behind in the care of grandparents. These Scots fisher-girls were as much a part of the Yarmouth fishing scene as the boats themselves.

Scots fisher-girls at work gutting herring. This photograph vividly shows the dirty and difficult conditions the girls worked under, out in the open whatever the weather, working from six in the morning often until late into the evening, depending on the size of the catches. The herring were tipped into the farlin, the long sloping wooden trough seen here, and lightly salted before being gutted. This process meant the girls quickly became splattered with herring guts, scales and blood. The work was piecework plus a small weekly wage, enough to pay board and lodging. A bonus for each filled barrel was paid at the end of the season. In 1931 and 1936, and again in 1949 and 1953, the girls went on strike for increased wages.

Many girls worked in the 'pickling yards' where the gutted herring were packed into barrels with layers of salt. When the barrel was full the cooper, in the centre of the group, would put the lid on.

This group of Scots 'girls' taking a break probably at one of the curing houses, shows the wide age range of the workforce. Regardless of age they were always referred to as 'girls' and, whatever the hardships, they always appeared happy.

From 1905 the Corporation tram system provided a public transport service from the Market Place to the Fishwharf, travelling via Blackfriars, Admiralty, Barrack and South Denes Roads, to a terminus near the old power station. Scots folk are seen here on the open top deck of a tram.

Motor buses took over from the trams on the Market Place to Fishwharf route in 1924. This bus, at the Fishwharf terminus, is one of three the Corporation bought from the London General Omnibus Co. in 1925. Note the solid tyres on this open top vehicle, which was in use until 1928.

Left: When not at work, groups of Scots fisher-girls would be seen walking around the town, busily knitting jerseys, or guernseys, for the fishermen. The ball of wool was usually kept in the apron pocket. Their strong Scottish accents made it difficult for local people to understand them.

Below: Fisher-girls, c.1900, seeking a welcome break at a portable coffee stall on the Fishwharf. Fish swills make do as seats for the older women, wrapped in shawls to keep out the cold.

Right: This postcard shows a Scots 'lassie' at work. At the height of the fishing season over 6,000 Scottish girls came to work in Yarmouth and Gorleston.

Below: An off-duty moment, in the 1930s, gives these three girls the opportunity to chat to a fellow countryman. The boat probably comes from their home port, and many girls were related to the men on the boats. Some met their future husbands during the fishing season. Boats from the Scottish ports of Buckie, Fraserburgh and Inverness are among those pictured here.

Empty swills provide seats for these off-duty girls as they busily knit and pose for the camera of the local press photographer in the 1950s. In the background is Ocean House, the head office of Bloomfield's Limited.

A few fisher-girls were local, and they usually worked in the curing houses. The majority of the female workforce, however, came from Scotland. Whether local or not they were all ready with a smile to have their picture taken when the opportunity arose.

Miss Davidson's Rest House for Scottish Fisher Folk was connected with the Hewett Fishing Company, a company involved in trawl fishing rather than herring drifting, but looked after the welfare of all Scottish folk in the port.

Nelson's Column is visible between the huge piles of empty barrels waiting for the girls to fill. Scenes such as this were common all over the South Denes, where most of the open land was used by the fishing industry for barrel storage, net drying or pickling plots.

Policemen try to clear a path and keep back the cheering crowds of fisher-girls as the Prince of Wales makes his way through on 21 October 1930. During this royal visit to the Fishwharf the Prince spent most of the morning being shown curing works and net chambers, meeting some of the girls and watching them at work. Note the bandaged fingers on many of the waving workers, whose enthusiasm is acknowledged by a raised hat from the Prince.

Opposite above: This postcard is dated 1913, the peak year of the industry. Many girls worked on the Gorleston side of the river, along the riverside north of the lifeboat sheds. Here the herring have been tipped into the farlins, ready for the girls to gut.

Opposite below: This postcard is dated 1910 and shows a line of girls gutting at the farlins while the coopers wait for the barrels to be filled. These highly skilled girls worked in teams of three, two gutting and one packing, starting work at 6 a.m. and often working 12 or 15 hours per day.

Above: Whether these girls are waving for the Prince, or for the cameraman, is not clear but they certainly enjoyed the special occasion of a royal visit. The Prince later went on to open the new Haven Bridge.

Left: These two girls are taking the gutted fish from the tubs ready to pack them into barrels. Each barrel held a cran of herring (about a thousand fish) and a hundredweight of salt. On 22 October 1936 the fisher-girls went on strike for a pay increase, from 10d to 1s (5p) a barrel. Millions of fish rotted on the quayside before the strike was settled four days later.

Opposite

Above: This team of girls are salting and packing herring at Bloomfields Ocean House pickling enclosure. Bloomfields were one of the largest firms in the fishing industry, owning a fleet of boats, curing yards, net works, ice plants and pickling plots.

Below: William J. Burton Ltd, fish curers and exporters, had premises in Admiralty Road, Middle Road East, Exmouth Road, Swanston's Road, Battery Road and Main Cross Road.

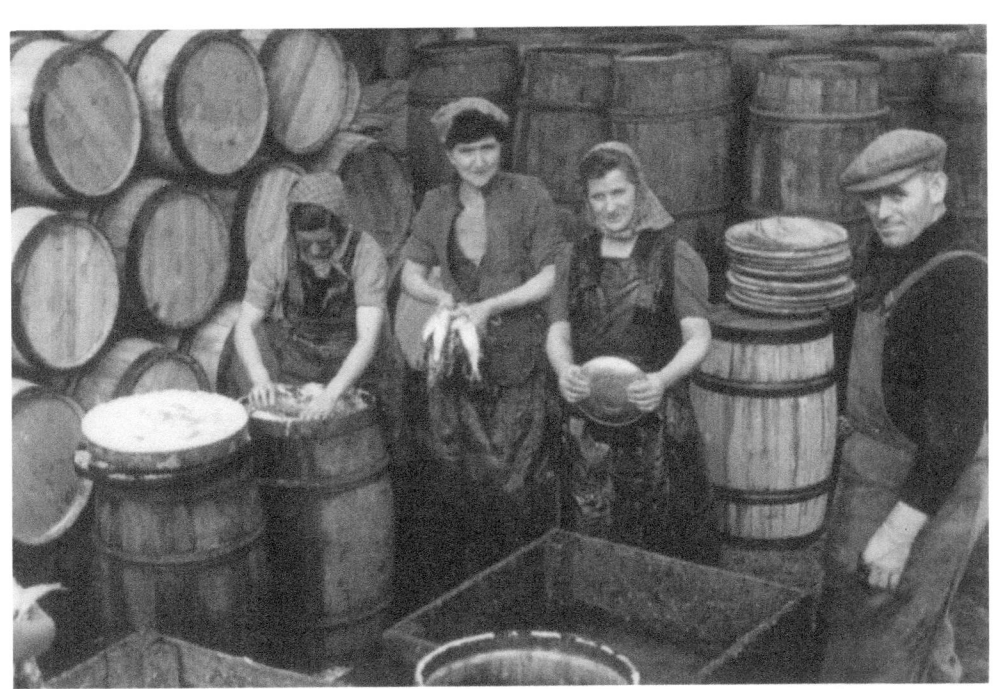

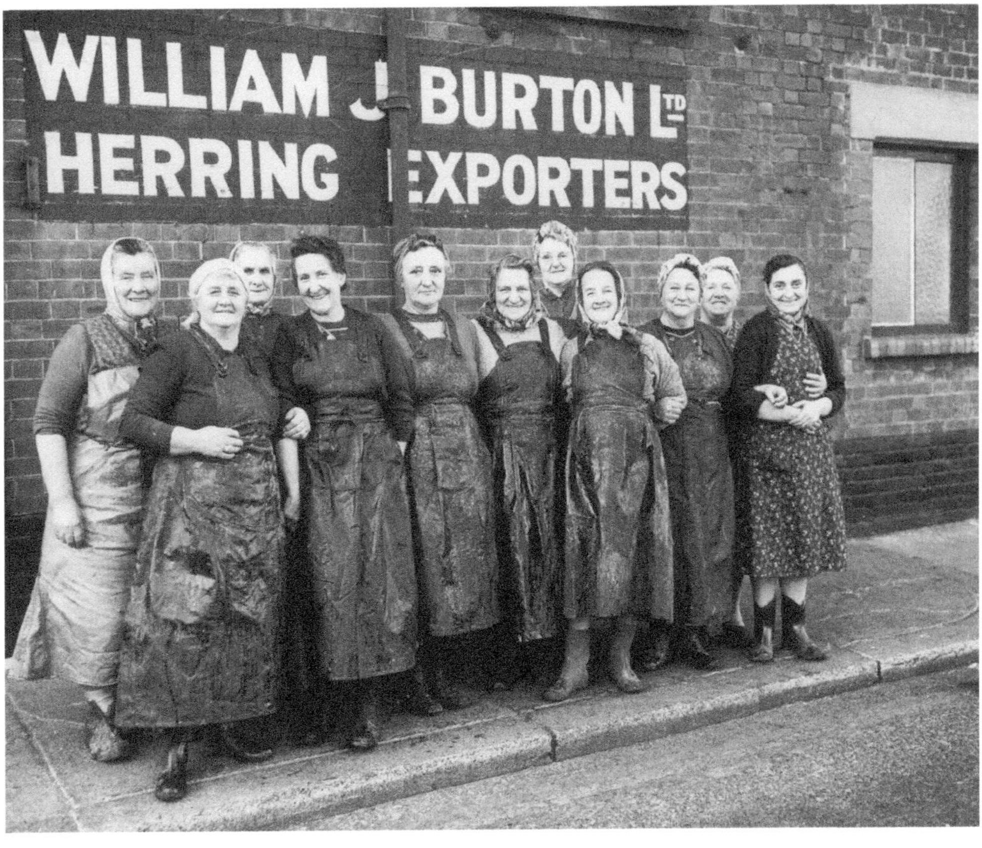
WILLIAM J. BURTON LTD
HERRING EXPORTERS

Packing on the South Denes in the 1950s. In the background are the Bloomfields works. Fish salted and packed in this manner, or pickled, was known as 'Scotch cure', a method of curing introduced by the Scots in the nineteenth century.

Opposite above: As the salt dissolved it formed a 'pickle' with the juices of the fish. After a few days the barrel had to be topped up with fish, and the liquid topped up with brine, before being finally sealed by the cooper.

Opposite below: Each morning the girls would wrap their fingers in cotton bandages, known as cloots, as a protection against salt sores and the sharp gutting knives. Many injuries still occurred, and these were treated at the dressing stations. Despite these hazards the girls were always cheerful, bringing the quayside to life with their laughing and singing.

SIX HIGHEST AWARDS.

Highest Award

Cured Herrings.

Special Prize.

Anti-Pilferage Package.

C. STACY-WATSON & CO. received the ONLY AWARDS at the National Fisheries Exhibition, Norwich, and the Fisheries Exhibition, Gt. Yarmouth, for HAM-CURED HERRINGS & ANTI-PILFERAGE PACKAGE, and the only Prize Medal awarded at the International Food Exhibition, London.

AND TWO DIPLOMAS OF HONOUR.

First Prize

Ham-cured, Export, and Pickled Herrings.

Special Prize

Herrings Preserved in Tins.

C. STACY-WATSON & CO.,
CURERS AND PACKERS OF THE
C_{ELEBRATED}

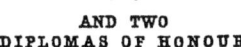

YARMOUTH HERRINGS

HAM CURE,

BLOATER CURE,

WHITE SALTED IN PICKLE,

REDS, ETC.

PACKED IN BARRELS, BOXES, AND TINS.

For Prices and Particulars address—

C. STACY-WATSON & CO., Yare Fishery Works, Gt. Yarmouth.

"THE SILVERY HOSTS OF THE NORTH SEA."
WITH HISTORICAL SKETCH OF QUAINT OLD YARMOUTH;
AND RECIPES
HOW TO COOK THE YARMOUTH HERRING.
By C. STACY-WATSON.
Price, cloth gilt, 1s. 6d.; coloured stiff covers, 1s.

LONDON: "HOME WORDS" OFFICE, 1, PATERNOSTER BUILDINGS, E.C.
GREAT YARMOUTH: C. STACY-WATSON.

The Yare Fishery Works of Stacy-Watson & Co., on South Denes Road, was one of the first large Victorian curing houses to be built on the open land at the south end of the town. Stacy-Watson's speciality was the 'Ham Cured Bloater', a breakfast delicacy, where the cured fish were packed into barrels and pressed down with a screw-press.

This view of the South Denes, taken c.1890, looking towards Gorleston, shows clearly the enormous number of barrels that would be used in one season. Considering each barrel would hold approximately one thousand fish it gives a good idea of the size of the industry in its peak years.

At work on a pickling plot. Every day a lorry would run on a circular route to the main dressing station, established by the Norfolk branch of the British Red Cross Society in St Peter's Road. The building is used today as the Red Cross Centre, and includes their charity shop.

When the film *The Kid Brother*, a silent comedy staring Harold Lloyd, was screened at the Empire in 1927, the manager Mr E. Bowles (at the wheel of the car) took the opportunity to advertise the film by organising a photo call at the Fishwharf. The horn rimmed spectacles and tiny straw hats were given to the fish workers as part of the advertisement, for which they probably received cut price entry to the cinema. The girls were known for their good humour and willingness to join in any occasion such as this. This film was regarded as one of Harold Lloyd's finest.

Opposite above: Many Scottish firms had depots at Yarmouth during the fishing season and would send both men and vehicles to the town. These men work for Sutherland's of Peterhead.

Opposite below: A Sutherland's lorry with staff from E&P Duncan, probably decorated for an end of season party before returning home for Christmas. Many of the girls, and the crews of the drifters, bought their Christmas presents in the town, before they left in late December. Shops stayed open late into the evening to benefit from this extra trade.

These nine girls probably comprise three separate teams. The packers, the three girls holding the shallow pans used to put the salt into the barrels, were usually in charge of the teams.

These fisher-girls work in a curing house. They are not wearing oilskin aprons. They are seen here on the back of one of Henry Sutton's lorries in 1959. Sutton's at this time claimed to be the largest firm worldwide in the smoked fish trade, exporting a million boxes of herring each year.

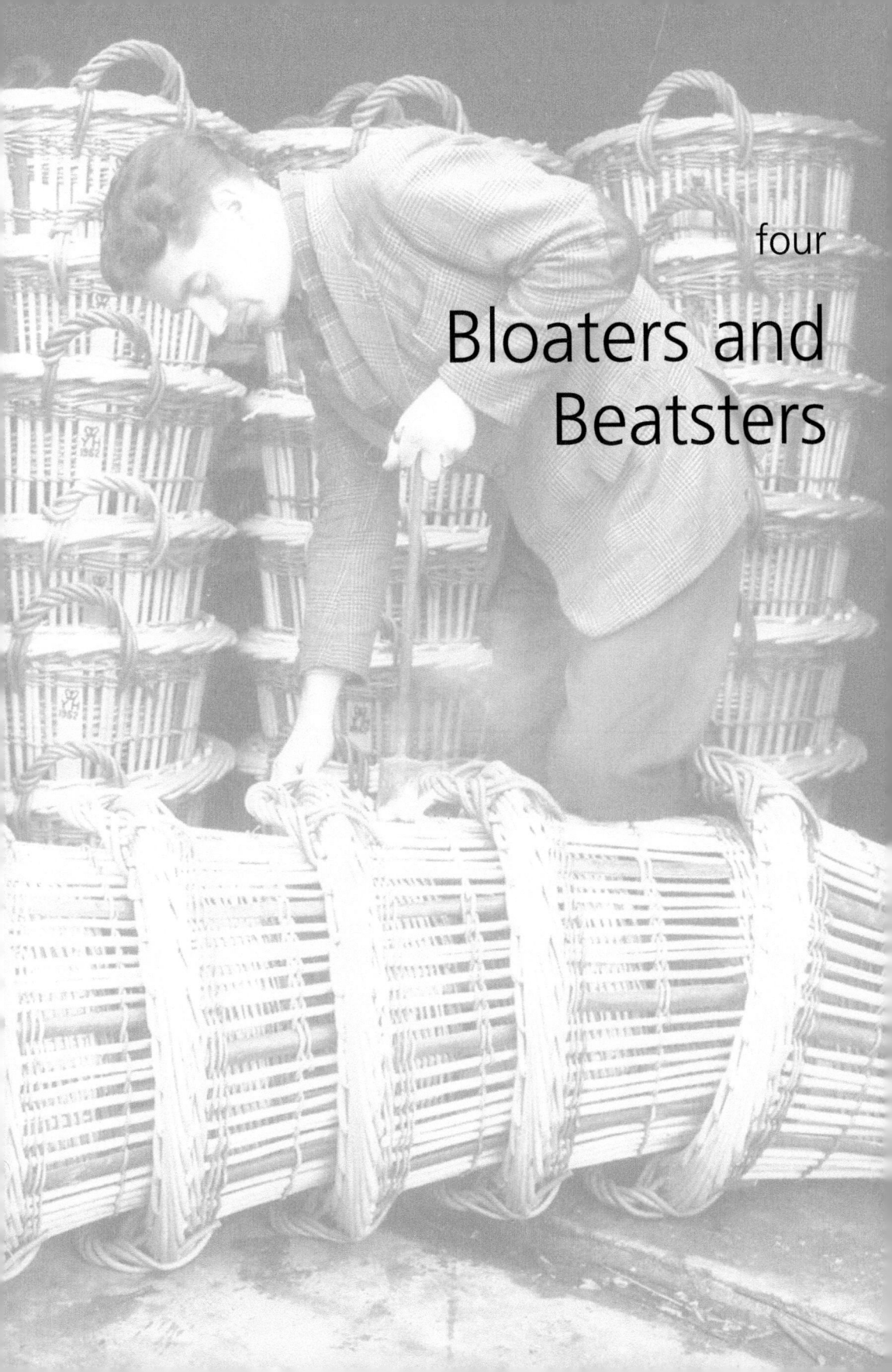

four

Bloaters and Beatsters

Large quantities of herring landed at Yarmouth were gutted and packed by the Scots girls into barrels, ready for export. This method of preserving the fish was known as the 'Scotch Cure', and introduced in the 1860s. The fish were tightly packed in the barrels between layers of salt, about one hundredweight of salt going into each barrel. After standing for eight days, the pickle, the dissolved salt and juices from the fish, was drained off and the barrel refilled with brine, topped up with fish, and sealed. Pickled herring were in great demand for the Russian and German markets.

Other herring were cured in the many smoke-houses that existed in the town. Fish-houses are recorded in Yarmouth from the thirteenth century. These early smoke-houses were built among the houses in the Rows. From the middle of the nineteenth century new, large, curing houses were built in the southern part of the town, on the undeveloped South Denes. At the curing house the herring were washed, dry-salted and graded before being placed in steeps, tanks full of brine pickle, where they were left for one to fourteen days, depending on the cure required. From the steeps the fish were 'rived', threaded onto thin sticks known as 'speets' which were pushed through one gill and out through the mouth of the fish, twenty to thirty fish on one speet. These were then arranged on racks in the smoke-house known as 'loves', the speets some six inches apart and one foot above each other, each chamber in the smoke-house holding up to seven lasts of herring. A fire of billets, wood shavings and oak dust was lit on the stone floor and allowed to smoulder, the smoke rising up through the fish and out through the well ventilated roof. When the cure was complete the fish were then ready for packing into barrels or boxes, either for the home market or for export. Throughout the fishing season a unique aroma hung over the southern part of the town, wafting from the curing house ventilators.

The 'bloater', a word synonymous with Yarmouth, was developed by a local curer named Bishop in 1836, the herring being lightly smoked, ungutted, for up to six hours. The bloater was a fish to be eaten as soon as possible after curing. Red herring, well known by the sixteenth century, were smoked for a week until the fish turned deep brownish red and these would keep for many weeks. The kipper, claimed to have been first produced by John Woodger of Newcastle in 1846, was a herring gutted and split open, before being smoked overnight. From 1918 some curers used a vegetable dye to produce a darker fish.

The vast quantity of nets required by the industry had to be constantly checked for damage and repaired in the many net chambers. These buildings were to be found in Gorleston and Caister as well as Great Yarmouth, and here the ransackers and beatsters worked. Ransackers were men responsible for checking the nets for damage before the women, the beatsters, mended them. Beating a net was a highly-skilled job, many girls starting to learn the trade at fourteen years of age. When the nets were completed they would be tanned, a process where the nets were boiled in cutch, in large coppers, a solution that not only turned them brown but waterproofed and protected them. Many other trades supported the fishing industry, skilled trades which included basket makers, box makers, cork cutters, coopers, mast and block makers and sail makers. There were also many engineering firms, ship chandlers and the suppliers of coal, ice and salt.

The herring were threaded onto long sticks called 'speets', each speet holding twenty to thirty fish, in a process known as 'riving'. The speets were then hung in the smoke chamber on the racks, or 'loves', before the fire was lit under them. A single smoke chamber could hold up to 7 lasts of herring.

After being smoked for the required period, depending on the type of cure required, the speets are placed on racks as seen here, ready for the packers. This photograph is dated 1955.

Gutted and split open, these herring are being prepared for the smoke-house to become kippers. The 'kipper speets' are wooden bars with rows of small hooks, on to which the fish are hung, ready for smoking.

KIPPERED HERRINGS!!

"The trade in **KIPPERED HERRINGS** originated at Newcastle-upon-Tyne in 1846, its founder being the late Mr. JOHN WOODGER. Introduced into London during the same year, the fish cured after a model method, were at first neglected, but before long made their way in public estimation, and became the subject of a great demand, which has continued to develop until in the present day there are about fifty firms engaged in the curing business. During the year, when the fisheries in England and Scotland are being actively prosecuted, these firms employ about 1,500 women and 300 men. The finest and freshest Herrings should always be, and are, by the best firms, used as Kippers. As affording some idea of the trade, we may mention that **Messrs. JOHN WOODGER and SONS send upwards of 8,000,000 Kippers to London and Country Markets alone yearly.** The Kippered Herring business was originated by this firm, and they are by far the **Largest Curers.**"
Extracted from "The Caterer & Refreshment Contractors' Gazette," Jan., 1880.

Our Cure now exceed 25,000,000 Kippers, the greater part of which are sold in London.

JOHN WOODGER and SONS, Ltd.,

Herring Curers and Merchants.

RETAIL SHOPS—

83, SOUTH QUAY,

AND

2, MARKET ROW,

GREAT YARMOUTH.

8, Baker St., Gorleston.

Branches:—Newcastle-on-Tyne, North Shields; Sea Houses, Northumberland, Hartlepool, Hull; Stornoway and Peterhead, Scotland.

The firm of John Woodger & Sons claimed to have been the first to produce a cure known as the kipper, in 1846. By 1912, when this advertisement appeared, they were producing twenty-five million kippers each year.

The prepared speets of kippers are hung in the smoke chamber. Below, on the brick floor, the fire consisted of billets, or wood shavings and oak dust, designed to smoulder and give off sufficient smoke and heat to cure the fish, but not cook it. The type of fire and level of ventilation depended on the cure required.

The cured fish were graded by size before being packed into boxes, some for the home market but most for export. Different markets required a different cure, these fish being packed for the Mediterranean.

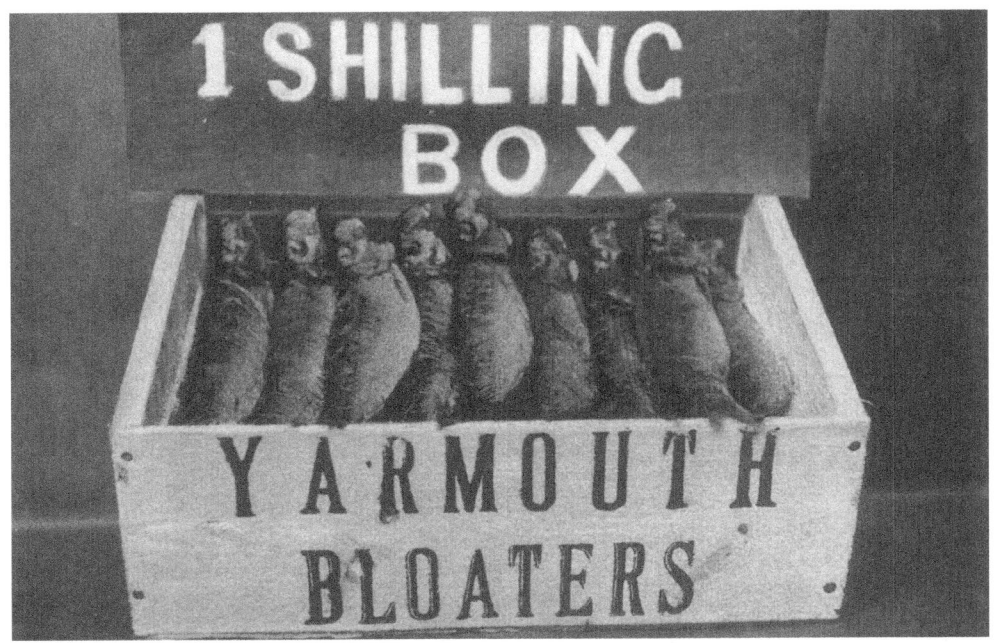

Yarmouth Bloaters were often featured on post cards available in the town during the holiday season, although this one was overprinted on the back as a Christmas card.

Today Mike Kelly, trading as HS Fishing 2000 Ltd, in what was originally part of Henry Sutton's smoke-house, carries out the only commercial curing in the town. Mike is seen here packing the cured herring for export. The fresh herring are imported from Norway and, once cured, they are exported to many different countries. Today, there is almost no home trade in cured herring.

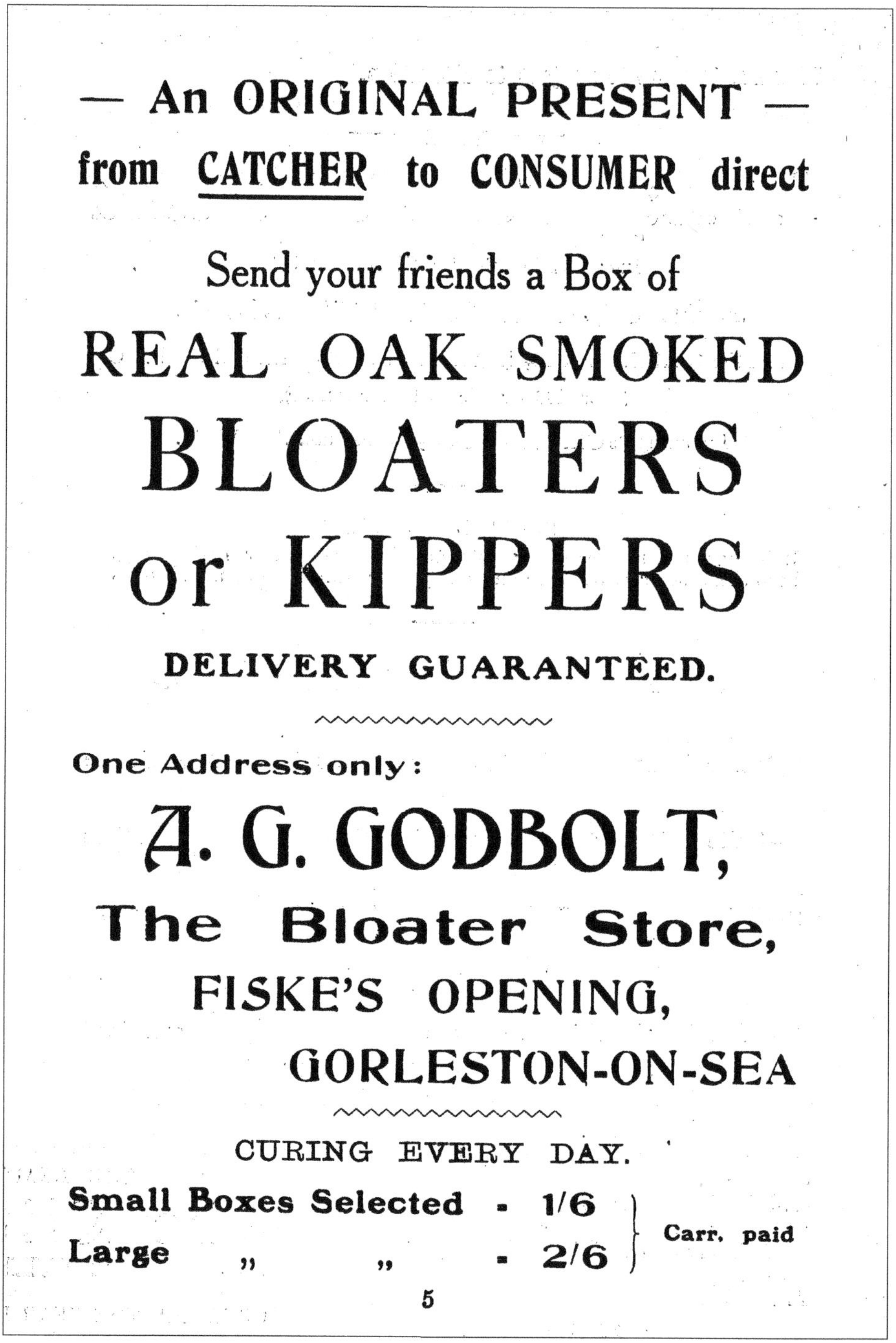

— An ORIGINAL PRESENT —
from CATCHER to CONSUMER direct

Send your friends a Box of

REAL OAK SMOKED
BLOATERS
or KIPPERS

DELIVERY GUARANTEED.

One Address only:

A. G. GODBOLT,
The Bloater Store,
FISKE'S OPENING,
GORLESTON-ON-SEA

CURING EVERY DAY.

Small Boxes Selected - 1/6 ⎫
Large   „   „   - 2/6 ⎭ Carr. paid

At one time there were more than sixty curing houses in Yarmouth and Gorleston. Some were very small, in the Rows, others were associated with fish shops or sometimes pubs, such as the Havelock Tavern, the Crystal Palace and the Blackfriars Tavern, all of which had adjoining curing houses.

Workers at the Co-op fish-canning factory in King Street in 1918. The factory closed in 1933 and had originally been Blanchflower's, where their famous Bloater Paste was manufactured. Potted meats and fish paste were very popular in the early twentieth century for afternoon tea.

Many attempts were made to mechanise the fish curing processes. Bloomfield's first introduced machines to bone herring in 1929. In the 1950s the machines seen here were being used at Bloomfields to gut herring, but it was still a very labour-intensive industry.

Ice was an essential commodity for the fishing industry and, before the introduction of machinery to produce artificial ice, the industry had to rely on a supply from natural sources. Each winter ice was collected from the frozen Broads and transported by wherry to this ice-house, where it could be stored for many months. In winters when the Broads did not freeze ice was imported from Scandinavia and stored here. This ice-house, originally one of a pair, adjacent to the Haven Bridge, was used to store ice until 1910. Today, although unused, this is the only surviving commercial ice-house of its type in the country and is now a listed building.

Opposite above: These blocks of ice have been produced by machinery at Bloomfield's ice factory in Bloomfield Road in the 1950s. The large blocks were then put through a crusher to produce ice of a suitable size for packing fish.

Opposite below: Many London fish salesmen had local agents working in the town on their behalf during the fishing season, as this advertisement from 1900 illustrates. William Symblett was also a boat owner.

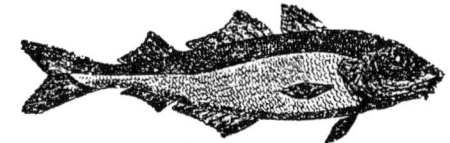

Telegraphic Address—"CRAPEFISH," LONDON.

EDWARD JEX & SON

FISH SALESMEN,

27, ST. MARY-AT-HILL

AND

23, Billingsgate Market, London, E.C.

☞ Cheques sent with accounts of Sales daily.

TRUNKS, BOXES AND TALLIES

Promptly forwarded on application to

E. JEX & SON, 27, St. Mary-at-Hill, London;

OR TO OUR AGENT,

W. H. SYMBLETT,

123, Havelock Road, Great Yarmouth.

BANKERS' REFERENCES—BARCLAY & Co., LIMITED, Lombard Street, London, E.C.

Baskets were used in great quantities by the fishing industry, until replaced by metal boxes in 1960. There were several basket makers in the town, the largest firm being Stanley Bird, making the traditional swills and quarter-cran baskets. The baskets were made from osier, a type of willow, most of which was grown locally on the Caister and Burgh Castle marshes.

These four basket makers, pictured early in the twentieth century, are making the quarter-cran baskets, used to unload the catch from the boat to the quayside. When finished each basket was stamped with a government mark, to certify it as an official measure.

Many local men were employed as coopers by the fishing industry and many more travelled down from Scotland. A cooper was expected to make up to eight barrels a day. In the 1930s the cooper was paid a fixed rate of 1s 3d (6p) per barrel.

Horace Johnson, a cooper, completing a herring barrel at Bloomfield's. Among the specialised tools used by the cooper can be seen a croze and a chiv. All the larger curing works had their own cooper's shops.

This Edwardian post card shows two off-duty fisher-girls, busily knitting while watching coopers at work making herring barrels. Postcards such as this were very popular with holidaymakers in the early twentieth century.

Thousands of barrels, stored around the pickling plots on the South Denes, with the girls and coopers working among them. Wenns Box Factory, on North Quay, a factory that also produced kipper boxes, supplied much of the wood from which the barrels were made.

Among the many industries associated with the fishing industry was that of sail making, particularly up to the end of the nineteenth century before the sailing drifters were replaced by steam drifters. This sail loft is a typical example of many once to be found in the town. By the 1930s there were only three sail lofts left in Yarmouth.

Making and repairing drifter nets was a large industry. Every season about 3,000 miles of nets were in use. With the nets went hundreds of ropes and thousands of corks. The men who checked the nets were known as 'ransackers' and the women who mended the nets were known as 'beatsters'.

Left: Paul Williment, a ransacker, checking the corks and 'norsels', the short lengths of twine holding the net to the head rope, in his net chamber at Gorleston in 1975. The building still exists on Riverside Road, and is now run by his son, Paul Williment junior, a ship chandler. It is the last net chamber in the town, and is almost in its original condition.

Below: Beatsters at work repairing nets in a Yarmouth net chamber. The nets are suspended from hooks in front of the large windows to give the workers the maximum light, the beatsters using a 'shale' or netting-needle to repair the mesh.

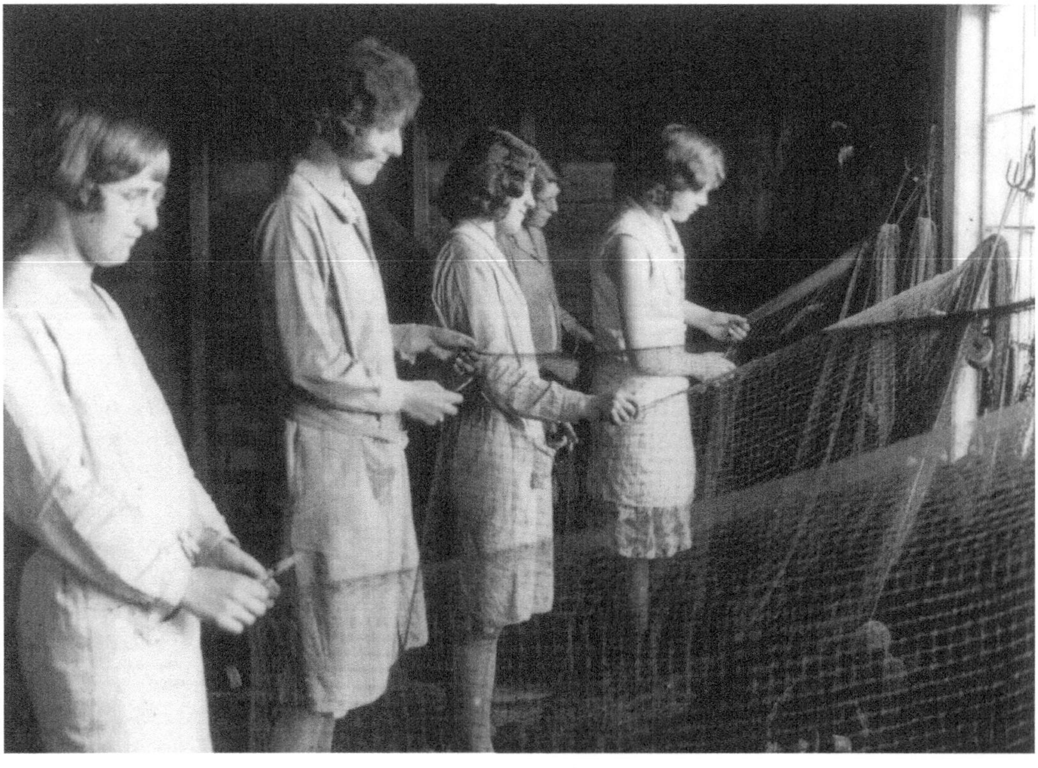

Looking down from the Monument on to part of the Bloomfield complex in 1946. In the centre of the picture is the Net Works on the corner of Admiralty Road and Salmon Road, with the net drying rails on the roof of the building. On the far left, on the corner with South Denes Road, is Ocean House, the main offices of the company.

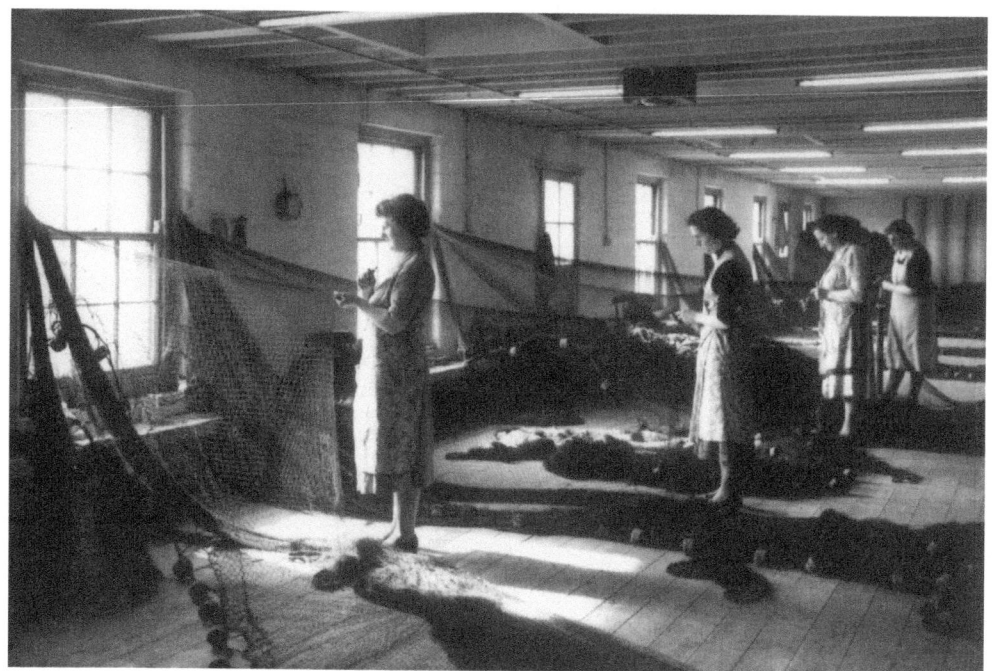

Beatsters usually worked on the upper floor of a net chamber, where there was more light. They mended all the broken or frayed meshes in the nets. Nets were easily damaged at sea and after each voyage many would need repairing.

A net chamber, this one being Haylett's at Caister, decorated for Christmas. From left to right: E. Brown, P. Haylett, L. Haylett, M. Larner, E. Symonds, C. Nichols, A. Dyble, T. George, A. Randell.

Above: In the small village of Caister there were nine net chambers, specially designed buildings, serving the Yarmouth fishing industry. Pictured here in 1968 is the Honeymoon Loke chamber, last used in 1949. At the far end of the building is the tanning copper in which the nets were boiled in cutch to waterproof them, and then hung over the rails to dry. The building is now a private house.

Right: Three Caister beatsters, from the left, -?- , Alice Bartley and Beatty Bartley. It took several years training to become an efficient beatster, girls starting to learn the skills required at the age of fourteen. The last Caister net chamber closed in 1961.

These steaming nets have just been removed from the tanning process, where they have been treated in a liquid containing cutch, a substance derived from the wood of an East Indian Betel-nut Palm tree.

Ransackers checking nets at Bloomfields net grounds off Salmon Road. Each boat required up to two miles of nets, so the total quantity of nets required by the fleets, including the spare sets, was considerable.

Above: To support the lines of nets in the water, large inflated canvas or plastic buoys, known as 'buffs', were fixed to the head rope. Some buffs were painted with owner's markings to identify them while some were left white.

Right: Yarmouth bloaters were known throughout the country, and Frederick Barnby was one of the first fish merchants in the town to advertise a service to send them to any part of the country. This advertisement is dated 1863.

YARMOUTH BLOATERS

Delivered free to any part of London, and to all Towns throughout England and Wales, where there is an authorised Agent. THESE FAR-FAMED FISH are only supplied by the undersigned during the best part of the season, from 1st October to the end of December.

12, 2s.,—25, 3s.,—50, 5s.,—100, 7s.,—200, 12s.,—400, 20s.,—1000, 35s.,—2000, 60s.,

Forwarded upon receipt of Post-office Order, Postage Stamps, or remittance through any branch of the National Provincial Bank of England.

FREDERICK BARNBY,
FISH MERCHANT,
GREAT YARMOUTH.
ESTABLISHED 1799.

High-Dried Herrings,
100, 7s.,—200, 12s.,—400, 20s.,

Are only supplied during the months of December, January, February, March, and April.

THESE FISH ARE CURED TO KEEP GOOD SIX MONTHS.

Hermetically Sealed, in Tins of 25 Fish, 10s. per 100, forwarded to all parts of the World.

FREDERICK BARNBY,
FISH MERCHANT,
GREAT YARMOUTH.
ESTABLISHED 1799.

Every year visitors sent home thousands of boxes of bloaters and kippers. Fishmongers would pack and dispatch the small wooden boxes to addresses all over the country. In the 1930s up to 5,000 boxes a day were sent by post.

On 23 July 1956, the Mayor, Mrs Gilham, visited the post office sorting office in Regent Street with Head Postmaster Mr Dunscombe, to see the boxes of fish posted by holidaymakers that day.

Above: This fishmonger, in South Market Road, also cured his own fish, as did many others. It traded under the unusual name of Frog's Hall, and was bought by the Co-op and demolished, c.1936. The large clock on the front of the building was later moved to new Co-op premises in South Market Road, the copper face having an unusual inscription, which read 'the moments passed, lay many fast'. It was said the clock was erected by the owner as a memorial to his wife, who was drowned in the local Suspension Bridge disaster of 1845.

Right: This advertisement by Mr Flertey, owner of Frog's Hall curing house, appeared in 1894.

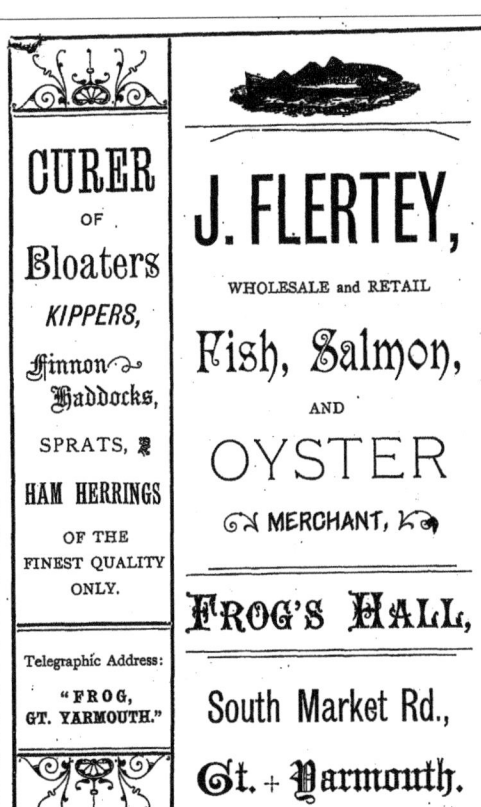

ALL WHO HAVE USED THEM SAY

Blanchflowers' Preparations

ARE UNEXCELLED FOR

BREAKFAST, LUNCHEON, TEA, &c.,

ON ACCOUNT OF THEIR

Delicious & Appetising Qualities.

THESE INCLUDE

BLANCHFLOWERS'
POTTED MEATS, FISH, & GAME,

Bloater and Anchovy Paste,

KIPPERED & HAM-CURED HERRINGS,

FRESH HERRINGS, SALT HERRINGS IN BRINE, HERRINGS IN TOMATO SAUCE,

DIGBY CHICKS, YARMOUTH BLOATERS,

Findon & Fresh Haddocks, Codfish, Turbot, Laitance de Hareng,

SAUCES, SAUSAGES, LUNCH & OX TONGUES, &c.,

PHYSICIANS COMMEND THEM FOR INVALIDS,

Connoisseurs, Cooks, & Confectioners pronounce them unrivalled for their

QUALITY, PURITY, AND ECONOMY.

Sold in Tins & Vases by all the leading Grocers, Store Keepers, & Italian Warehousemen.

J. C. BLANCHFLOWER & SONS,

Proprietors and Manufacturers,

GREAT YARMOUTH, ENGLAND.

THE GREAT YARMOUTH PRINTING Co. LTD LITHO

Timothy Coleman Blanchflower filed a patent in 1868 for his Bloater and Meat Pastes. This advertisement shows the wide range of products the firm was soon producing, advertised as 'Table Delioacils' [sic]. Blanchflowers' factory was in King Street, producing delicacies for the Victorian table. Digby Chicks are similar to Yarmouth Bloaters, but produced from Atlantic, rather than the North Sea herring.

The demand for rope and twine was considerable, and at one time the industry employed 150 people in the town. John Taylor Bracey began as a rope maker in 1802, in Deneside, the firm later moving to Exmouth Road, pictured here, c.1900. Bracey's were the first firm in the town to install machinery for rope making. By 1927 they were listed as net makers, closing in 1939. The site, at No.1 Exmouth Road, later became the factory of the Yarmouth Cardboard Box Company.

Ropewalks were often a quarter of a mile long. A bunch of fibres, known as a 'head', were attached to a hook on a spinning jenny, which revolved as the spinner walked backwards along the walk. This produced a strand, three or more strands then being twisted together to form a rope. In 1904 there was a ropewalk along the western side of Ordnance Road.

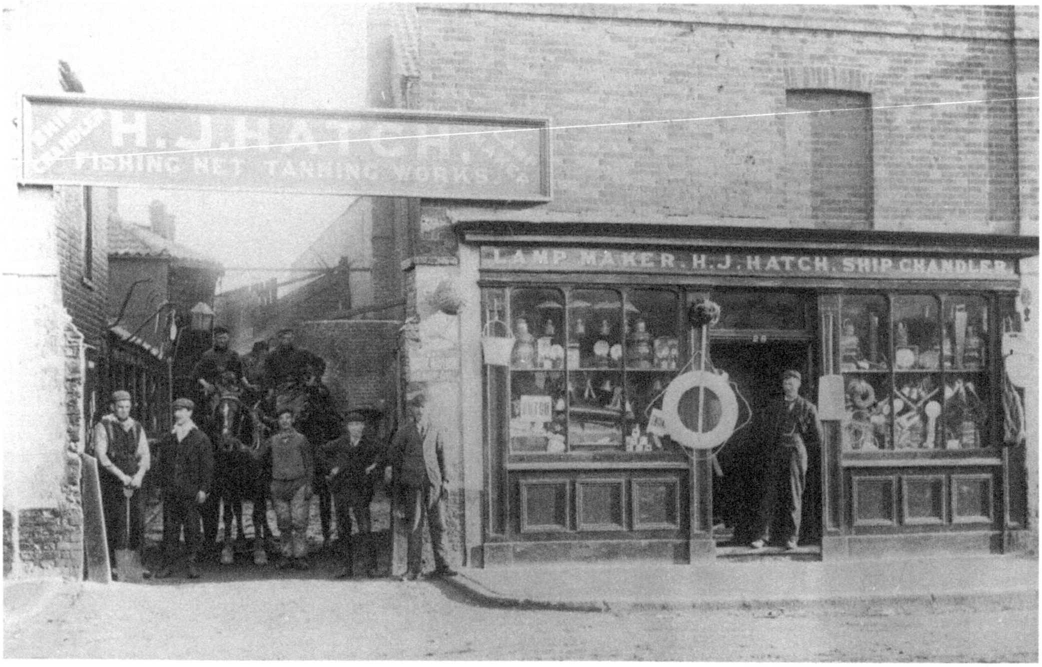

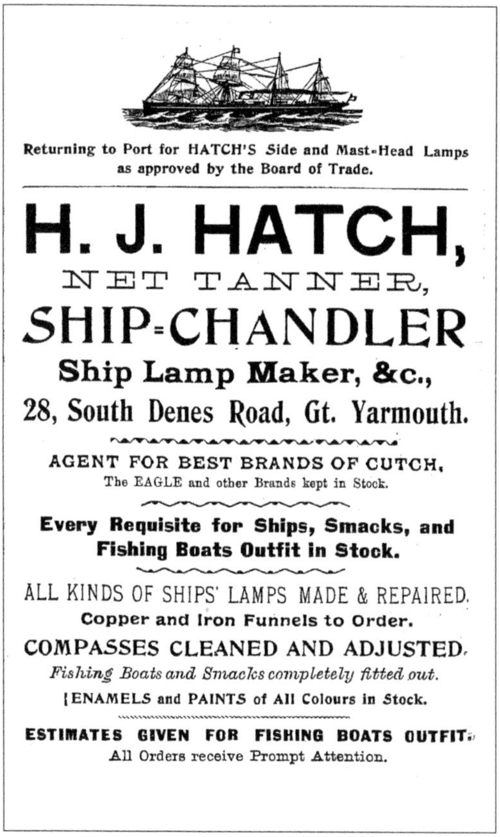

Above: The staff of H.J. Hatch pose for a photograph outside their premises in South Denes Road. The ship chandler was able to supply boats with almost every requirement. The other large ship chandler in the town was Yarmouth Stores Ltd.

Left: Hatch's advertisement shows just some of the vast range of items a ship chandler was able to supply.

Returning to Port for HATCH'S Side and Mast=Head Lamps as approved by the Board of Trade.

H. J. HATCH,
NET TANNER,
SHIP=CHANDLER
Ship Lamp Maker, &c.,
28, South Denes Road, Gt. Yarmouth.

AGENT FOR BEST BRANDS OF CUTCH,
The EAGLE and other Brands kept in Stock.

Every Requisite for Ships, Smacks, and Fishing Boats Outfit in Stock.

ALL KINDS OF SHIPS' LAMPS MADE & REPAIRED.
Copper and Iron Funnels to Order.
COMPASSES CLEANED AND ADJUSTED.
Fishing Boats and Smacks completely fitted out.
ENAMELS and PAINTS of All Colours in Stock.
ESTIMATES GIVEN FOR FISHING BOATS OUTFIT.
All Orders receive Prompt Attention.

In 1952 a Herring Reduction factory was built by the Herring Industry Board on land towards the harbour mouth, at a cost of £100,000, to convert surplus herring into oil, fertilizer and cattle food. It was the first factory of its kind in the United Kingdom, equipped to process 800 cran per day.

Fish that nobody wanted. A Fraserburgh drifter moored at the reduction factory, where many thousands of fish are piled on the quayside awaiting processing. However, the expected glut of herring never materialised and the factory closed after a few years, as did the whole industry a few years later.

Other titles published by The History Press

Fishing from the Humber
ARTHUR G. CREDLAND

This book is a unique pictorial history of Hull's fishing industry and its community; from the role of the trawlers during the two World Wars to the amalgamations of the 1960s and 70s, the infamous Cod Wars and the massive reduction of the fleet after 1975. Despite the dark clouds, there has been good news, and October 2001 saw the opening of the new £4.5 million Hull fish market, 'Fishgate', a new chapter in Hull's fishing heritage.

0 7524 2813 6

Great Yarmouth and Gorleston Pubs
COLIN TOOKE

This comprehensive volume of archive images recalls the intriguing history of many of Great Yarmouth and Gorleston's pubs, some of them still trading, others long since closed or demolished. With 200 illustrations, this outstanding collection records the many varied roles pubs have played in the social life of the area. From Lacon's Brewery, pub outings, horse drays and maltings, to evocative advertisements from the nineteenth century, each image offers an insight into the popularity and changing role of local pubs.

0 7524 3298 2

Herring A History of the silver darlings
MIKE SMYLIE

The history of the herring and those whose lives have revolved around taking it and getting it to the tables of the masses. The book looks at the effects of the herring on the people who caught them, the unique ways of life, the superstition of the fisher folk, their boats and the communities who lived for the silver darlings. For those who've neglected the silver darlings for lesser fish such as cod and haddock, there are numerous recipes to try.

0 7524 2988 4

An Eye on the Coast The Fishing Industry from Wick to Whitby
GLORIA WILSON

This collection of photographs and drawings is a personal celebration of fisher people, harbours and boats used along the coast of Scotland and parts of the North East. These pictures are not intended to form a complete record of vessels and fishing communities, but rather, to show an appreciation of the classic wooden-hulled, cruiser-sterned seine netters and dual-purpose craft. Many of these splendid boats have perished under decommissioning schemes of recent years and so this book could be seen as a lament to a passing era and will undoubtedly evoke many strong memories.

0 7524 3853 0

If you are interested in purchasing other books published by The History Press, or in case you have difficulty finding any of our books in your local bookshop, you can also place orders directly through our website

www.thehistorypress.co.uk